Geoscience Information

A state-of-the art review

W0111520

Proceedings of the 1st International Conference
on Geological Information
London, 10 – 12 April, 1978

Edited by

Anthony P. Harvey & Judith A. Diment

THE BROAD OAK PRESS LTD.,
Publishers
Heathfield, Sussex, England

The Publisher's colophon is the 'King of the Woods' oak,
Jedburgh, Borders, Scotland
from **A history of British trees** by P.J. Selby (1842)

First published in 1979 by
The Broad Oak Press Limited, Ragstones, Broad Oak,
Heathfield, East Sussex, England TN21 8UD

ISBN-13: 978-0-906716-00-7   e-ISBN-13: 978-94-011-7444-2
DOI: 10.1007/978-94-011-7444-2

CONTENTS

Application of information handling to applied geology

Documentation in specialized areas

User viewpoints

INTRODUCTION

The International Conference on Geological Information represents the first major attempt to bring together geoscience information specialists from all over the world.

The purpose of the conference was to assess the current state-of-the-art in geoscience information from both the regional and functional point of view. It was hoped that the conference could take steps to bring about increased international cooperation and collaboration in the field of geological information. The papers ranged over the whole spectrum of documentation from primary publishing back to the user, including data. Perhaps a keyword for the conference might be "cooperation". The idea of, and need for, cooperation was stressed in almost every talk. The final panel session was devoted to a discussion on the formation of a proposed International Association for Geological Information. Despite the growing pressure on information managers, stimulated by increasing international activities in geology, the global perspective of plate tectonics and worldwide concern for the availability of non-renewable resources, there does not exist an international organisation specifically concerned with geological information. Delegates agreed that there was no need for a new professional society of individuals but that a federation or similar organisation might be desirable. In the final session it became apparent that if the geological information community is to make the best use of all the systems and developments available there is very clearly a need to know what exists in all these areas at present. An urgent task is to identify these systems. There is also a need to look outside the geological sphere to examine the various national policies for science information and the various systems of national and international control to see how they effect geology and to impress upon them the particular needs of geologists.

An ad-hoc committee was established with representation from all the organisations involved in sponsorship of the conference including the Geological Information Group of the Geological Society of London, the Geoscience Information Society, Australian Geoscience Information Association, Editerra (European Association of Earth Science Editors) and AESE (Association of Earth Science Editors).

The committee will arrange a second international conference on geological documentation and it aims to compile a report for that meeting which will entail a world survey of geological information resources and a recommendation on the feasibility and desirability of the formation of a Federation of Geological Information Organisations.

THE EDITORS

iii

STATE-OF-THE-ART IN GEOSCIENCE INFORMATION - USA

DEDERICK C. WARD

Earth Science Librarian

University of Colorado Libraries, Boulder, Colorado 80309

Summary: In the 1960s, US geoscientists developed a pattern of diverse
and, for the most part, unrelated science information files, chiefly
for internal use by public and private organizations. When Federal funds
to sustain information operations ended up in the 1970s, producers of
information learned to survive under real market conditions and respond
to a new user community of regional planners, policy makers, and a public
concerned about energy and the environment. Although certain indicators
show that the geoscience information community is moving to meet new
demands, neither geoscience nor the nation's science establishment as
a whole share information responsibilities in a formalized way. Dramatic
changes in the traditional view of scientific and technical communication
have spawned an information industry, and geoscientists will not escape
its impact.

1. BACKGROUND

In the mid-1960s, US geoscientists, working under Federal subsidy,

joined their counterparts in the other sciences to study the information

problem and its impact on communication in the earth sciences. In

response to this initiative, the Geoscience Information Society was

founded in 1965 to deal with the issues created by an expanding body

of science literature, and, in 1967, the President of the American Geo-

logical Institute (AGI) appointed a committee on Geoscience Information

to plan a national geoscience information programme. The plan was never

implemented, but, as discussed later in this paper, there are signs

that an interest remains in a network concept.

The thinking of those active in geoscience information in the late

sixties and early seventies is summarized in two reports. A concept of

an information system for the geosciences, American Geological Institute

(1970), described a pattern of diverse and, for the most part, unrelated

science information files developed for internal use by special user

groups both in public and private organizations. Against a background of

rising operational costs, the report suggested that these separate units

join in a national information system to share the responsibilities of

geoscience information in as compatible a way as possible.

The second report took the form of a testimony submitted to the

National Commission of Libraries and Information Science by the Geoscience

Information Society (Bichteler, 1974). Like the 1970 report, the lack

of a geoscience information system in the United States was a major con-

cern, but the 1974 report set the problem more clearly against the needs
of the user and the unique multi-disciplinary character of geoscience
documentation.

2.   CURRENT STATE-OF-THE-ART

The current state-of-the-art in geoscience information can be considered a
response to basic changes in the economy and the user community.

2.a.   IMPACT OF THE ECONOMY

US information personnel now describe, as practical, an information service
which can survive under competition in a restrained economy and independent
of government subsidy. As Federal funds to sustain information services
became unavailable, different ways have been devised to use our technology
to produce information less expensively and to make it more quickly
available to the appropriate user. Some examples will illustrate this
trend.

In the first presidential address devoted entirely to communication
in geology, Gordon Craig (1969:305) said, "In the near future Editorial
Boards will have to decide whether a paper should be retained in type-
script form, or what parts should be published in a journal, photographed
on microfilm or card, or stored in a data bank." A decade later, the
Geological Society of America (GSA), faced with a backlog of manuscripts of approximately 30 months, steadily lengthening delays in publishing, and a 14% annual rate of increase in costs, will issue its
Bulletin articles in microfiche, retaining the traditional paper Bulletin
issues for the publication of summaries. GSA adopted this new format
with the conviction that the production of high-quality work, like the
science of geology itself, is growing, not shrinking, and that increased
selectivity in accepting papers would not solve the problem. This new
format will allow the publication of approximately 80 papers per issue,
at decreased costs, and with 2 to 4 months between the acceptance of camera-
ready articles and publication.

Since its inception in 1967, a most significant innovation of the
American Geological Institute's GeoRef automated citation file has been
public access to the file via remote terminals on-line with the System
Development Corporation's computer in California. AGI issued a thesaurus
to the file in 1977 (American Geological Institute, 1977).

Consideration of ways to organize stored data for special user groups
has prompted AGI to explore, with the consumer, the production of special

bibliographies, such as the Bibliography and index of micropaleontology
and the new Mineral exploration abstracts, bibliographies of geology of
individual states or regions (Colorado, Kansas, the Gulf Coast), and
bibliographies produced as current-awareness features in specialized
journals (Marine Geology, Engineering Geology, Geoderma).

Although showing a 100% increase in total royalty in 1977, the
GeoRef file is still markedly under-used by American geologists. However,
with the complete transfer of the Bibliography and index of geology from
the Geological Society of America to AGI on January 1, 1979, and the combined bibliographic services on a sound financial base for the first time,
the Institute can now turn its attention to marketing and seek ways to
educate the geologist to use a computer to retrieve information.

GeoRef is not the only AGI bibliographic service under-used by
American geologists. Following a period (1959-1970) of subsidy by the
US National Science Foundation, coupled with a loss of subscription revenues
chiefly from academic institutions, the AGI Translations Program reduced
its publishing from six translation journals to a single one - the
International Geology Review.

A policy of selection has kept the International Geology Review
economically viable for 19 years, and it is generally felt that this
journal plus those presently covered by the translation programs of
Plenum Publishing Company, Scripta Publishing Company, and the American
Geophysical Union (AGU) provide the geoscientist with the best articles
from the Chinese and Russian literatures, given the current level of
support by the profession. The AGI Director of Publications suggests
that a demand for further coverage of geoscience literature in translation
will be tempered by the limits dictated by basic economics.

In 1974, the Committee on Data Interchange and Data Centers of the
US Geophysics Research Board recommended a study be undertaken regarding
new demands that are being placed on World Data Centers and associated
national data centers by large-scale geophysics programs. Exponential
growth, once solely an issue with published literature, is clearly a
problem with numerical data. Geophysics programs are supplying data
several orders of magnitude greater in information content than the old
types of data, and they are appearing on magnetic tape in one of a variety
of digital (or sometimes analog) format. In a report published by the
National Research Council in 1976, one of the recommendations of a geophysics study panel was that the cost tradeoffs between acquisition, handling, and archiving of data be considered an integral part of planning, and
that thinning and compression of data be considered. The panel suggested
that a significant part of the cost of depositing data in data centers in
an accessible form be borne by the programmes that generate those data.

The realities of economy foster innovation. In the period from
May 1973 to December 1976, members of the geoscience community - notably
the US Geological Survey - conducted a series of test conferences using
a computer as conference chair. The computer conferences were held
first among US participants; later the conferences became international.
Exploiting such computer qualities as memory and standardization, computer
conferencing promised the following advantages (Vallee, Askevold, & Wilson,
1977:3):

(1) More deliberate answers to technical questions, backed up by
facts and with less delay;

(2) More effective coordination of technical projects among geographically separated participants;

(3) Improved ability to deal with a large volume of information;

(4) Improved ability to introduce human judgement in an information
system.

Finally, a statement should be made concerning the impact of economics on the American college campus. Faculty and students on university
campuses have been slow in adapting automated library services to their
own use, even when such services are readily available. Dwindling funds
to match soaring book costs and a reluctance by administrators to plan
new library buildings, however, are powerful factors which will force
academicians to research their information using new technologies and
reading more in micro-format. Another factor favouring the introduction
of new information technology on the campus is the promise of moderation
in the ever-increasing size of annual requests to legislatures and other
funding agencies for the funding of traditional library materials and
support staffs.

2b. A CHANGING USER COMMUNITY

A second basic change that has affected the thinking and activities of
those agencies producing geoscience information has come as a result
of a decided shift in the number and type of users requesting the information. A shift in emphasis from the advancement of science per se to
public concern over environmental, ecological, and energy issues has
created a demand for scientific and technical information in a format
which can be understood by economists, citizens' groups, business people,
lawmakers, attorneys, and regional planners (Austin, 1977).

Fortunately, there are indicators which show that the geoscience
community is moving to meet the demand. In 1970 the US Geological Survey
began the first of seven urban area studies in cooperation with the US
Department of Housing and Urban Development - the San Francisco Regional

Environmental and Resource Planning Study. The goal of the study was
to provide both basic and interpreted earth science information needed
for regional planning and decision-making. The study, now completed,
will result in publication of a series of over 90 geologic, hydrogeologic,
and topographic maps, technical reports, and interpretive reports - all
designed specifically for the 107 distinct city and county user groups
in the area (Kockelman, 1977). The Survey continues to issue, through its
Office of Environmental Geology and Energy Lands Program, reports and maps
in a form understandable to planners and policy makers. Preliminary
reports, once available only through inspection at certain national depos-
itories, can now be purchased by mail in either microfiche or hard copy.

The production of environmental geologic atlasses and related
publications is not limited to the Federal government. In 1979, a significant number of land-use and geologic hazard studies were published
by approximately 12 state geological surveys (Passero, 1978).

Realization that government agencies are spending millions of
dollars for information files servicing a restricted number of users
has not escaped the managers of numerical data systems. The question
they face is what proportion of observational, processed, and interpretive data will satisfy new user requirements. National Oceanic and
Atmospheric Administration's (NOAA) Environmental Data Service (EDS),
originally established to provide traditional archiving and dissemination
services, is involved in adapting its data to help those seeking solutions
to national and global environmental problems. A current example of
this trend is the establishment, by EDS, of a network of Regional Coastal
Information Centers to meet the needs of state, county, and local
officials and business groups engaged in activities associated with the
coastal zone (Austin, 1977:6).

The Petroleum Data System (PDS), maintained by the University of
Oklahoma under contract to the US Geological Survey, is a group of
data bases which include information on over 75,000 oil and gas fields
in the US and Canada. Initially designed for purposes of reserve and
resource analysis within the Geological Survey, the file is now available
to the public through the General Electric time-sharing network (Meyer,
1975: Tracy, 1978). Both PDS and its citation-companion at the University
of Tulsa, the Petroleum Abstracts Information Service anticipate expansion
of their files to include more complete coverage of energy resources and
economic minerals - thus creating a complete energy information network.

3. NETWORK CONCEPT

In a paper presented to the Geoscience Information Society in 1974, T.M. Albert, Director of the National Energy Information Center, Washington, D.C., wrote, "In the Federal Government alone, more than 43 agencies carrying on more than 350 separate programs collect some form of energy-oriented statistics. The quantity of individual items of numerical data collected is staggering. No one has taken an inventory, but FEA (Federal Energy Administration) has made a beginning and is proposing to do more. The amount of duplication is probably quite extensive" (Albert, 1976:45).

Even as there is progress in adjusting to the economic and political realities of the 1970s, there remains the formidable and persistent problem of duplication of effort, and the lack of that unity and purpose as could be best expressed by an information network. Neither geoscience nor the nation's science community as a whole share information responsibilities in a formalized way. At present, the only standing national committee representing geology's information interests is AGI's GeoRef Advisory Committee. This leaves the rest of US geoscience information activity without direction. A national geoscience information system (American Geological Institute, 1970), as conceived by proponents in the early 1970s has not materialized, due, no doubt, to the lack of initiative from the profession's own ranks following the withdrawal of the government planning subsidy.

Interest in cooperative efforts is not totally dead. The government's own Oceanic and Atmospheric Science Information Service (OASIS) and its data-companion ENDEX (Environmental Data Index) access commercial information systems rather than duplicate their files and services (Austin, 1977:5). The beginning of increased cooperation in geology could result from a recently funded proposal by the American Geological Institute to design an experimental network for sharing information and data resources by linking five state geological surveys, the US Geological Survey, and AGI (American Geological Institute, no date). This study will highlight the political, organizational, and economic problems that are likely to be encountered in a cooperative venture of this kind and test the concept of networking as a means for information resource sharing among relatively small, specialized research communities.

4. INFORMATION AS AN INDUSTRY

The demands of economy, the emergence of new information markets consisting of unique user groups, and the bringing together of those who participate in the profession (eg, the International Conference on Geological Information, London, 1978) have all contributed to the fact that information has evolved from a library-oriented, locally-conceived science to an industry which involves marketing, selling, legal problems, competition between public and private services, and political problems arising from competition between information services of different nations. These dramatic changes in the nature of information raise important questions concerning ownership and conditions of access to the information.

It is the impact of this new industry upon traditional communication practices that summarizes the current state-of-the-art in US science and geoscience information. The balance between public and corporate responsibilities and the sharing of information between those who manage it and those who need it are the topics which need resolution, but there is little doubt that the traditional views of scientific and technical communication have changed and that the geoscientist is intimately involved in that change.

Acknowledgments: I am grateful to the following colleagues for discussing with me current developments in geoscience information: George Becraft, Office of Mineral Resources, US Geological Survey; C.F. Burk, Canada Centre for Geoscience Data; Herb Meyers, Solid-Earth and Marine Geophysical Data Division, National Geophysical and Solar-Terrestrial Data Center; John Mulvihill and Thomas Rafter, Jr., American Geological Institute; Jean Thyfault, Geological Society of America.

REFERENCES

ALBERT, T.M. 1976. Technical information programs of the Federal Energy Administration. Proceedings. Geoscience Information Society 6 : 44-51.

AMERICAN GEOLOGICAL INSTITUTE. 1970. A concept of an information system for the geosciences. Falls Church, Virginia, 23pp.

AMERICAN GEOLOGICAL INSTITUTE. 1977. GeoRef thesaurus and guide to indexing. 1st ed. Falls Church, Virginia, 391pp.

AMERICAN GEOLOGICAL INSTITUTE. No date. Design of an experimental cooperative network for sharing information and data resources in geology; a proposal submitted to the National Science Foundation. Falls Church, Virginia (not paged, mimeographed).

AUSTIN, T.B. 1977. Challenges in meeting public needs for science information. EDS (Environmental Data Service Newsletter) November 1977 : 2-6.

BICHTELER, J. Editor. 1974. Geoscience information and user needs; testimony submitted to the National Commission on Libraries and Information Science by the Geoscience Information Society. 26pp. (Summarized in Geotimes 19 : 29, June 1974).

CRAIG, G.Y. 1969. Communication in geology. Scottish Journal of Geology 5 : 305-321.

KOCKELMAN, W.J. 1977. Use of earth-science information by city and county planning agencies in the San Francisco Bay Region, California. Proceedings. Geoscience Information Society 7 : 45-55.

MEYER, R.F. 1975. U.S. Department of the Interior - Energy Data Files Proceedings. Geoscience Information Society 5 ; 89-95.

NATIONAL RESEARCH COUNCIL, GEOPHYSICS RESEARCH BOARD, GEOPHYSICAL DATA PANEL. 1976. Geophysical data centers: impact of data-intensive programs. Washington: National Academy of Sciences, 32pp.

PASSERO, R.N. 1978. Environmental geology. Geotimes 23, no. 1 : 25-26.

TRACY, P.A. 1978. Petroleum Data System - A network of energy information. Proceedings. Geoscience Information Society 8 : 25-30.

VALLEE, J., ASKEVOLD, G., & WILSON, T. 1977. Computer conferencing in the geosciences; a report prepared for the US Geological Survey. Menlo Park, California: Institute for the Future, 85pp.

CANADA: CURRENT ACTIVITIES AND ISSUES IN GEOLOGICAL

DOCUMENTATION

CORNELIUS F. BURK, Jr.

Canada Centre for Geoscience Data

Department of Energy, Mines and Resources,
Ottawa, Canada, K1A 0E4

Summary: With the world's second-largest landmass, Canada
faces a formidable challenge in maintaining documentation
services for the geological sciences and the resource
industries. In addition to publications, other important
sources of documentary information include public government
files, theses, and specialized data services, the latter
mainly in support of the petroleum industry. A national file
for secondary (bibliographic) information has been built
cooperatively by federal and provincial agencies, with
emphasis on unpublished information, and a national referral
system has been partially implemented. Among important issues
affecting geological documentation are the lack of effective
national policy for scientific and technical information (STI)
in general, including lack of bibliographic control for
Canadian-produced and Canadian-related information, and the
impact of geological information on public policy issues such
as non-renewable resources and the disposal of radioactive
wastes.

Sources of geological information

Primary literature: The major publishers of geological
information in Canada are the federal government, the
professional societies, provincial government agencies and
various trade journals related to the mining and petroleum
industries. Among federal publishers, the Geological Survey
of Canada and the National Research Council of Canada are
probably the best known. Professional societies active in
publishing include the Geological Association of Canada, the
Canadian Society of Petroleum Geologists, the Mineralogical
Association of Canada and the Canadian Institute of Mining and
Metallurgy. Nearly all of Canada's 10 provinces publish
geological information through their departments of natural
resources or research councils. Examples of trade journals
are Oilweek, the Northern Miner and the Canadian Mining
Journal.

Government unpublished information: A substantial volume of
unpublished (but public) information is maintained by most
federal and provincial agencies in Canada. The Geological
Survey of Canada and the Mineral Policy Sector of the
Department of Energy, Mines and Resources and the Department
of Indian and Northern Affairs maintain various open files and

mineral inventory records. Provincial government agencies maintain files of assessment reports submitted by mining and petroleum companies. In addition, resource agencies maintain extensive, usually comprehensive collections of borehole logs, well data files, cores and samples. The province of Alberta has been a leader and pioneer in preserving such basic geoscience data for the benefit of future research and exploration.

Universities: Canada has more than 25 graduate schools in geology, geophysics and engineering which produce geological information in the form of M.Sc. and Ph.D. theses. Some of this information is later published.

Specialized data services: Mainly in support of the petroleum industry in western Canada, over 20 private companies provide specialized data services dealing with exploration, production and engineering. The Alberta government is presently consolidating and improving its capabilities to acquire and manage basic data related to petroleum, coal and oil sands through development of the Energy Resources Data System (ERDS) (Morin and Ethier, in press). The project is expected to take 5 years to complete at a cost of an estimated Cdn. $3.6 million.

Other specialized data services include, for example, those providing remote sensing data and the Geological Survey of Canada which routinely places digital data obtained from geochemical and geophysical surveys on open file; several organizations are in the process of developing mineral deposit files (e.g. Picklyk, 1976; Longe, 1978).

Secondary services

Canada is developing a national secondary file, the Canadian Index to Geoscience Data, and the country also draws heavily on the international services, particularly GeoRef of the American Geological Institute and Geoarchive of Geosystems, United Kingdom.

The Canadian Index to Geoscience Data has been developed over the past 10 years by the Canada Centre for Geoscience Data and its predecessors, in collaboration with Canadian government agencies at both the federal and provincial level (Gunn and Burk, 1975). The purpose of the file has been to consolidate identification and indexing of documents containing information on the Canadian landmass and its offshore regions. Twelve contributing agencies have indexed more than 65,000 titles, about 70 per cent of which are unpublished reports. The principal method of public access will be through an on-line service bureau, beginning in 1979. In addition, the contributing agencies have published various indexes based on the national file (e.g. Gregory, 1978; Poruks and Hamilton, 1976; Groen, 1978). A national professional society has independently produced a similar index to its publications (Hay and Dixon, 1978).

Policy and operational procedures for management of the Index are the responsibility of the Canada Centre for Geoscience Data, with support from the national Geoscience Index Advisory

Committee. The Centre has recently rejoined the multilingual thesaurus project of the Committee on Geological Documentation, International Union of Geological Sciences.

Canadian scientists have access to the GeoRef file produced by the American Geological Institute through the national selective dissemination of information service (CAN/SDI) provided by the Canada Institute for Scientific and Technical Information and through on-line facilities offered by Infomart. Various other databases of interest to geoscientists are also available on-line (Silcoff, 1978).

Coverage by the above-mentioned secondary services can be viewed in terms of "on Canada", "from Canada" and "for Canadians". Coverage of information on Canada is probably about 70 per cent complete. The extent of coverage of information from Canada is not known, but certainly it is incomplete, since there is no operational program to establish bibliographic control of Canadian-produced geological information. The Canada Centre for Geoscience Data is currently examining the possibilities of developing such control. Coverage by the secondary services of information required to meet the overall needs of Canadian geologists is not known but it is clearly incomplete.

National information programs

Information programs of national scope in Canada that relate to the geological sciences include development of general policies for scientific and technical information (STI), programs devoted expressly to geology and finally, national information programs in related areas.

A general policy for STI was advanced by the Science Council of Canada in 1969 (Katz, 1969) and shortly thereafter adopted by the federal government. Responsibility for developing specific policies rests with the Advisory Board on Scientific and Technical Information (ABSTI) which reports to the National Research Council of Canada. Operationally, the national focus for STI in Canada is the Canada Institute for Scientific and Technical Information (CISTI, 1977). To date, neither of these bodies has had a direct impact on information policy for system development within geological programs of the federal or provincial governments, or on the geological profession as a whole; there is no overall national system framework for STI within which geological information systems could be developed.

Nevertheless, geologists in Canada have been pioneers in the development of broadly-based information systems. Development of the so-called National System for Geological Data recommended in 1967 (Brisbin and Ediger, 1967) has evolved into a national referral system, which shows promise of maturing over the next few years as a fully functional national information system. The objective of the Canadian Geoscience Data Referral System (CGDRS), as it is presently called, is to provide knowledge on the existence, location and nature of public geoscience data dealing with the Canadian landmass and its offshore regions. The system concept includes elements to cover policy, international liaison,

capture and acquisition of secondary data, national
coordination and operations, the generation and use of
referral tools and, finally, the user community (Burk, 1978).

CGDRS is similar in concept to the model for scientific and
technical information recently adopted in the Federal Republic
of Germany (Institute for Documentation, 1976).

National information programs in related areas include those
operated by the Canada Centre for Remote Sensing (McGurrin and
Silcoff, 1976) and a national referral service provided by the
Technology Information Division of the Canada Centre for
Mineral and Energy Technology (CANMET) (Romaniuk, 1977; Taylor
and Kanasy, 1978); another national activity of interest
relates to the development of guidelines and standards for
mineral deposit data (Longe, 1978).

Issues

In summary, four major issues face those responsible for
geological documentation in Canada:

1.   Financing the primary literature: In common with
      publishers throughout the world in all fields, the
      financing of journals, technical reports and other
      primary literature in Canada is a growing problem.
      Although information technology has enhanced our ability
      to process and disseminate some kinds of information, it
      has not produced comparable effects in the handling of
      primary literature. Costs continue to escalate and the
      market remains static or becomes smaller.

2.   National STI Policy: Current and future development of
      national information systems in support of geologically-
      based activities in Canada is vulnerable due to the
      absence of a general systems framework for scientific and
      technical information (STI). As the investment of time,
      money and other resources increases with the passage of
      time, the potential risk in losing this investment
      through bad planning and lack of coordination becomes
      greater.

3.   Bibliographic Control: There is no nationally organized
      program to identify and control Canadian-produced or
      Canada-related geological information. Responsibility
      for this task has not been assigned or assumed, due in
      part to the lack of information policy. The solution
      will undoubtedly depend on cooperation through some
      appropriate mechanism with the international
      bibliographic services and other external information
      organizations.

4.   Impact of Geological Information on Public Issues:
      Geological information and supporting information
      management programs have value to the extent that they
      improve decision-making for national programs, issues and
      goals. Current public issues in Canada with important
      geoscience components include, for example, energy and
      mineral resources and the development of sites for
      radioactive waste disposal. It is in the best interests

of the geological profession and its supporting
information services that the communication of the
geological aspects of these issues to scientists,
legislators and the public at large be carried out more
effectively than has been the case in the past.

References

BRISBIN, W.C. and EDIGER, N.M. Editors 1967. A national system
for storage and retrieval of geological data in Canada.
National Advisory Committee on Research
in the Geological Sciences, 175pp.

BURK, C.F. Jr. 1978. The national data referral system for
Canadian geoscience. Geoscience Information Society
Proceedings 1977, 8 : 31-41.

CISTI 1977. Canada Institute for Scientific and Technical
Information report 1974-77. National Research Council
of Canada 16014, 42pp.

GREGORY, D.J. 1978. Index to open file reports. Nova
Scotia Department of Mines, Rept. 78-3, 39pp.

GROEN, H.A. 1978. Index to published reports and maps,
Division of Mines, 1891 to 1977. Ontario Geological
Survey Misc. Paper 77, 408pp.

GUNN, K.L. and BURK, C.F. Jr. 1975. The Canadian index to
geoscience data: a decentralized, cooperative indexing
project. Canadian Association for Information Science
Proceedings 3d Conference, Québec, 242-253.

HAY, P.W. and DIXON, C.R. 1978. Subject-author - NTS area
index. Canadian Society of Petroleum Geologists, 470pp.

INSTITUTE FOR DOCUMENTATION. 1976. The programme of the
federal government [of West Germany] for the promotion of
information and documentation (I & D - programme).
Institute for Documentation (IDW), Frankfurt a.M., 125pp.

KATZ, L. Chairman. 1969. A policy for scientific and
technical information dissemination. Science Council of
Canada Rept. 6, 33pp.

LONGE, R.V. Chairman. 1978. Computer-based files on mineral
deposits: guidelines and recommended standards for data
content. Geological Survey of Canada Paper 78-26, 72pp.

McGURRIN, B. and SILCOFF, B. 1976. Remote sensing information
centre, a national information network and node.
Canadian Association for Information Science Proceedings
4th Conference, London.

MORIN, E.J. and ETHIER, J.M. (in press). Alberta energy
resources data: past, present and future. In MIALL,
A.D. Editor. Facts and principles of oil occurrence.
Calgary: Canadian Society of Petroleum Geologists,
Memoir.

PICKLYK, D.D. 1976. An index to Canadian mineral occurrences
- preliminary considerations. Computers & Geosciences 2:
317-319.

PORUKS, M. and HAMILTON, W.N. 1976. Index to uranium
assessment reports for quartz mineral exploration
permits, northeastern Alberta. Alberta Research Council
Rept. 76-6, 81pp.

ROMANIUK, A.S. 1977. Making better use of existing
technological information. Northern Miner 63/37 : D8-9.

SILCOFF, B. 1978. Databases available to Canadian users
through six computer citation retrieval network centres -
as of March, 1978. Canadian Journal of Information
Science 3 : 110-122.

TAYLOR, G.W. and KANASY, J. 1978. Special challenges and
problems in a technical information service serving the
mineral and energy industries. Canadian Association
for Information Science Proceedings 6th Conference,
Montreal, 197-211.

GEOLOGICAL DOCUMENTATION IN THE FEDERAL REPUBLIC OF GERMANY

HARM GLASHOFF

Documentation service,
Federal Institute for Geosciences and Natural Resources,
P.O.B. 510153, D-3000 Hannover 51

Summary: A brief report on the state of the art of geological documentation in West Germany related to the general structure of scientific research, the information programme of the government, and the user's market.

The state and development of geological documentation in West Germany is best understood when related to the general structure of both scientific reserach and documentation in the country.

1. General research situation.

Research is carried out mainly by the universities, which are subordinated without exception to the regional states, by federal and state agencies, and to a lesser degree by industrial research centres; one difficulty arises from the federal constitution, which gives the responsibility for culture and science (including scientific and technical information and documentation) to the regional states. In practice these difficulties are minimized by the activities of research foundations (as Deutsche Forschungsgemeinschaft and Max-Planck-Gesellschaft), and by agreements between regional and federal governments.

2. The Organization of Information and Documentation.

In 1962, the federal government pointed out the political and economic value of scientific and technical information -

this was one year before the famous "Weinberg Report" of the US government was edited. Since then it has worked to establish an overall plan to sponsor information and documentation activities, initially through specialized institutes, such as the well known "Institut für Dokumentationswesen". The plan was finally presented by the federal ministry of research and technology, and was accepted in 1974. The aim of the plan is to concentrate the widespread activities of some 500 existing documentation centres into 16 major centres by discipline (such as medicine, physics, etc.), and 4 by scope (standards, patents, environmental problems, on-going grant aided research), and to concentrate the information and documentation infrastructure to a new founded "Gesellschaft für Information and Dokumentation". These 20 centres are planned to form a national network within the EURONET, called "Online Dokumentations- und Informations-Netz (ODIN)".

The organizational form of the centres will vary from commercial to public agency according to their historical development and financial structure.

3. The user market in Geology.

Some 30 universities offer geological training. They employ approximately 500 geologists and educate 5,000 students (1975). The eleven geological surveys and agencies employ about 450 geologists, and the federal survey 220. Together with grant aided employees and some 800 geoscientists in industry, the total number of specialized users can be estimated to be about 7,000 to 10,000. The number of non-geologists potential users can be assessed as 10% of this figure. These figures are supported by the use made of the existing information service.

4. The state of geological documentation.

In 1969 the board of directors of the geological surveys agreed to implement a geological documentation centre at the federal survey at Hannover. In the initial stage in 1970 the Hannover group joined with the on-going development at the Bureau de Recherches Géologiques et Minières in Orléans. This project was merged later with the geological documentation of the Centre National de Recherches Scientifiques at Paris

(Bulletin Signalétique).

Because of the language barriers between French and German geologists, the system designed was bilingual and made use of computer translation programmes. Information with respect to library holdings in the two countries (library codes and shelf numbers) was also merged. According to a contract between the German and French partners, the centre in Hannover supplies citations to the geological literature issued in the Federal Republic, the Democratic Republic, and Austria. For use in Germany the regional literature is indexed more deeply (e.g. topographical map sheets, regional descriptors, mineral names) than exchange version. This is similar to the computer internal "Bibliographie de Geologie de France" of the Bureau de Recherches Géologiques et Minières.

The German literature is stored since 1970 and the exchanged international literature from 1973, amounting to some 220,000 titles. The bibliographic data base is available to the public through special user oriented programmes, which were developed together with the mentioned translation and merge programmes by the Bundesanstalt für Geowissenschaften und Rohstoffe in loose cooperation with SIEMENS; but the programmes are fully compatible to IBM 360/370 machines.

5. The use of the documentation centre.

The number of requests in 1978 was nearly 2,000. Although no specific publicity was made until 1978, 40% of the requests came from universities, 10% from industry (this number is now increasing), 10% from individuals, and 40% from the state and federal surveys.

Special output programmes are available for the creation of catalogue cards, those in DIN-format as the most preferred, or in international library format, as listings or as tape for further processing. The regional geological citations are processed further for photocomposition in the "Zentralblatt für Geologie, Teil 1".

References:

THE PRESIDENT'S SCIENCE ADVISORY COMMITTEE. 1963. Science, Government, and Information. The responsibilities of the technical community and the government in the transfer of information. Washington: US Government Printing Office. 52 pp.

BUNDESMINISTER FÜR FORSCHUNG UND TECHNOLOGIE. 1975. Programm der Bundesregierung zur Förderung der Information und Dokumentation (IuD-Programm) 1974-1977. Bonn: Bundesminister für Forschung und Technologie. 147 pp.

GLASHOFF,H. & SCHOLZ,R.W. 1975. Literaturinformation und -dokumentation in den Geowissenschaften. Mitteilungen aus dem Geologisch-Paläontologischen Institut der Universität Hamburg. 44: 343-352.

GEOLOGICAL INFORMATION IN FRANCE

J. GRAVESTEIJN

Documentation Department

Bureau de Recherches Geologiques et Minières, Orléans (France)

Summary : In France, the Bureau National de l'Information Scientifique et Technique coordinates and promotes national information activities in the various scientific and technological fields.

Support has been given to the development of networks, multilingual tools, on-line searching, installation of data bases on host computers and the coordination between the national network and EURONET.

The major geological information service in France is operated by the Bureau de Recherches Géologiques et Minières through its Geological Survey and the Centre National de la Recherche Scientifique through its Centre de Documentation Scientifique et Technique.

The common geological data base PASCAL-GEODE is available for on-line searching through ESA/SDS in Frascati.

The oil industry has always played an important part in the development of the information sciences in France. The various oil companies started and developed the on-line activities in France.

1 - BUREAU NATIONAL DE L'INFORMATION SCIENTIFIQUE ET TECHNIQUE (BNIST)

In France, the BNIST was created by the government in 1973 in order to coordinate and promote information activities in science and technology. The BNIST programme of action concentrates on the following objectives :

- National and international networks : integration of existing networks or creation of new networks in fields where duplication exists. For example, in the field of chemistry, an agreement has been signed between the French chemical information associations and Chemical Abstracts Services and in the textile sector, an international information system TITUS (Traitement de l'Information Textile Universelle et Sélective) is now operational. In this system the content of the documents is defined by concepts connected by syntaxic relations using intermediate language. Abstracts written in this metalanguage can be translated automatically. In the atomic energy field, national information activity has been integrated in International Nuclear Information System(INIS) and there is a national cooperation program in which producers and users participate.

- Development of multilingual information tools : the BNIST supports work on the multilingual thesaurus for geology, the multilingual aspects

of TITUS and gives research grants for the automatic translation of
abstracts.

- On-line processing : several contracts have been signed for the
development of on-line software ; management of networks, natural
language searching, management of thesauri, etc.

- Coordination between the national information network and EURONET :
work has been carried out for the definition of the data bases to be
installed on French host computers which will be connected to EURONET.

- Special programs and contracts for coordination of scientific
editing : studies have been made on production costs and distribution
problems for the French scientific primary journals and on the position
and impact of French scientific publications in foreign countries, etc.

In the field of the earth sciences, BNIST support has been concen-
trated on the development of multilingual information tools (Bilingual
indexes, multilingual thesaurus project of the International Union of
Geological Sciences (IUGS) and the International Council of Scientific
Unions - Abstracting Board (ICSU AB).

2 - PRIMARY JOURNALS

The French primary journals are in a difficult position since the
market is small and most of them have less than 1000 subscribers. Most
of the earth science journals are published by learned societies or
public institutions such as universities, Bureau de Recherches Géologiques et Minières (BRGM), Office de Recherche Scientifique et Technique
Outre-Mer (ORSTOM), etc. In some cases, geology is grouped with other
scientific fields so that the journal can have a wider audience.

Some statistics can be given. In France there are 200 journals
which may publish geological papers. Of these 130 are geoscience journals
(30 of a regional type and 100 are devoted to special fields such as
micropalaeontology, petrology, mineralogy, etc), 70 journals are general
scientific periodicals containing some geological papers (Comptes rendus
de l'Académie des sciences, Bulletin du Muséum d'Histoire naturelle, etc).
Of this category 20 are "national" journals and 50 are of a regional kind.
The total number of geological papers published annually in France is
about 3000.

3 - GEOLOGICAL INFORMATION SERVICES

The geological documentation and information services were created
just before or during the Second World War. The Centre for Scientific
Information of the Centre National de la Recherche Scientifique (CNRS)

was set up in 1939 and the first issues of the Bulletin Signalétique including the earth sciences were published at that time.

The information activities of the BRGM also date back to the same period. The service began as a paleontological data file but the scope was soon widened and in 1950 a complete bibliographic card system was operational. Based upon a classification using a subject heading code, this card system continued until 1968. In that year, BRGM started the first computer-processed geological bibliography Bibliographie des sciences de la Terre with the automatic output of citations, author's indexes, permuted subject indexes and geographical indexes. The system was based upon in-depth indexing using a very elaborate thesaurus.

In 1972, CNRS and BRGM came to an agreement for the joint publication of the bibliography called Bulletin signalétique - Bibliographie des sciences de la Terre. The principles were shared indexing and abstracting with a common policy for marketing, pricing and subscription management.

In recent years, cooperation has been further extended and now concerns not only the production of the common data base PASCAL-GEODE and the printed bibliography but also the sale of magnetic tapes and on-line searching.

The PASCAL-GEODE data base is available via European Space Agency - Space Documentation Service in Frascati. Batch processed selective dissemination services and retrospective searches are offered separately through the PASCAL-VIRA system for CNRS and through the GEODE system for BRGM.

The French information centers have always played an active part in international cooperation, both for scientific and economic reasons. The best service from the end user's point of view would be a completely comprehensive system. Obviously, this can only be attained through international cooperation in which each partner handles his own national documentation. Sharing of the indexing work is also the best means for reducing the cost of production per item.

Since 1968, cooperation agreements have been signed between the French centers and foreign institutions. In most cases, cooperation is conducted through the national geological survey and in this respect cooperation programs exist between BRGM-CNRS and Bundesanstalt für Geowissenschaften und Rohstoffe in Hannover, GEOFOND-Prague, Empresa Nacional Adaro-Instituto Geologico y Minero de España, Madrid, the geological surveys of Hungary, Poland, Rumania and Finland.

Cooperation is based upon the exchange of services. The national geological surveys, through their information departments make total or partial analysis of the geological information published in the country. In exchange, the French partner sends copies of the printed version of the bibliography or the tapes containing the complete world bibliography. Both sides profit from the relationship.

Indexing is based upon the common thesaurus which has been completely translated into English, German, Spanish and Italian and partially into Finnish, Hungarian, Polish and Rumanian.

This European geological information network operates in a satisfactory way but there are, of course, practical problems. In some cases, delays in indexing may occur but the analysis is often speeded up by the fact that the national centre can analyse galley proofs of primary publications (Gravesteijn, 1974)

BRGM and CNRS also take an active part in international organizations such as ISCU AB, the European Association of Earth Science Editors-EDITERRA, International Standard Organization (ISO) and contribute to the United Nations World Science Information System (UNISIST) programs.

The information centres of the Institut Français du Pétrole (IFP) (the Research Centre of the Oil Industry) and the Société Nationale ELF-Aquitaine (SNEA) did much pioneer work in the field of scientific and technological information. For several decades, IFP has published bibliographic cards indexed according to a coded classification scheme. The information covers various aspects of petroleum research and technology. Both IFP and SNEA were among the first institutions to be connected to System Development Corporation, Lockheed and the European Space Agency and contributed largely to the expansion of on-line searching in France.

On-Line searching is not yet fully accepted by the scientific community. Financial, linguistic and training problems as well as old habits are the major restricting factors.

The BNIST promotes the installation of terminals in the university libraries and covers the investments. At present, about 10 universities are connected to the telecommunications networks.

4 - ASSOCIATIONS

The Société Géologique de France has no special information group and therefore most of the geological information specialists and librarians are members of the Association Française des Documentalistes et Bibliothécaires Spécialisés (ADBS) or the Association Nationale de la

Recherche Technique (ANRT). These associations organize professional discussions, clubs, training sessions, seminars and a biannual national information congress which is usually attended by 400 - 500 persons. They are also involved in the definition of standards and set up working groups for the study of special aspects of the profession such as the role of the documentalist in the world of science.

During the last 3-4 years regional subgroups have been created so as to promote participation of the individual documentalist in professional activities.

5 - CONCLUSION

In France as in all other developed or developing countries information is not yet considered by the end user as a basic raw material. However, the most recent technological developments such as on-line searching are changing this picture. The users are benefiting from improved access to information and we observe an increase in the use of bibliographic data bases.

This trend will no doubt continue but we should remember that while computer technology has made enormous progress, information handling has lagged behind.

In France as well as in many other countries there are still bottlenecks in the different stages of the information flow.

In the future the information services and data base producers will have to create or further develop compatible tools for guiding the end user in his search and the primary literature will have to be made available to the end user in a more efficient way.

REFERENCES

BUREAU NATIONAL DE L'INFORMATION SCIENTIFIQUE
1975. Rapport annuel d'activité. Paris : BNIST
1976. Rapport annuel d'activité. Paris : BNIST
1977. Rapport annuel d'activité. Paris : BNIST

GRAVESTEIJN, J. 1974. Geological Information - An example of international cooperation. In : Proceedings of the ICSU AB General assembly meeting in Berlin. Paris : ICSU AB : 221-233.

STATE-OF-THE-ART OF GEOLOGICAL DOCUMENTATION
IN THE UNITED KINGDOM

GAVIN H. BROWN

Brown's Geological Information Service Ltd.,

160 North Gower Street, London NW1 2ND

Summary: Geological documentation in the UK is undertaken by a variety of agencies - government, learned society, commercial and academic. Nearly all these agencies lie across the neat lines of this classification however. This wealth of industry and talent has itself been documented recently by the Geological Information Group in a compendium of geological information resources in the UK. Barriers to improved documentation activity include lack of funds, but a stronger incentive at official level towards improved documentary services is one desirable outcome of this meeting.

The speaker for the host country has the advantage that many of those engaged in these activities are present and able to speak for themselves during the meeting. Furthermore our host organization the Geological Information Group of the Geological Society has only just recently compiled a directory of geological information resources in Britain, which will be published in the next weeks (Diment 1978), and to which those interested may refer for a detailed review of organizations engaged in geoscience documentation in this country.

This paper will therefore attempt merely to sketch in the background to this activity, in the hope of illuminating its special characteristics. One of them as Professor Sutton has mentioned is the lack of centralization in policy making; but there are grounds for supposing that this country - despite its shortcomings - may claim to stand as a model for the future direction of geoscience documentation activities.

The preoccupations of those who originate and manipulate documents in geoscience can be viewed at three levels: there is the basic information:
- what is known about the geology itself
- where is the information
- who is compiling it
- in what form is it recorded.

There is a second tier, consisting of interest in special forms of document:
- maps, monographs, data files, indexes, theses, serials, conferences, translations, photographs, computer systems;

and of special interests: subject and area studies.

At a third level superimposed on these are the technical problems:

- rate of advance of knowledge
- structure and politics of geological organizations
- cost of recording and publishing

and so on.

One suspects that the preoccupations of this meeting are likely to be with finding technological solutions to increased and increasing:

- rates of publication
- costs of recording and publishing data
- diversity of sources and sinks of information.

But one could hope that it will also address itself to some value judgements, including the ones mentioned below, since it seems fair to say that technological solutions merely juggle with the symptoms of problems, never coming to grips with real causes.

Some issues perhaps need particular attention:

- criteria for selectivity in documentation
- criteria for discrimination in what to record/collect/publish
- means for the integration of documentation activities - at different levels of specialization and at different stages down the line of communication.

Our criteria for selectivity must obviously include common sense: we don't want articles published twice; we don't want articles which add little or nothing to our knowledge; and also we don't want articles which are incomprehensible. These are three different problems, all of which depend upon proper refereeing. Yet they are, because of the publish-or-perish syndrome, familiar ones. Only by facing up to those who publish at any cost will we bring this cancer under control; we should aim therefore, not merely for completeness of coverage in our information systems, but for recognition of merit. We now have advanced technology which is theoretically capable of weeding out whatever is unsatisfactory after it has got into the system; eventually this may prove to be one of the best justifications for the use of computers to handle geological information. But however we do it, we have to cleanse the channels of communication of an overload of half-digested information.

In Britain some relevant work is being done on the topic of discrimination in what is published. M. D. Gordon of the Primary Communications Research Centre at Leicester University presented a study at the First International Conference of Scientific Editors (Gordon 1978) in which work is described which has brought to light clear evidence of subconscious bias among referees of scientific papers. Preference was demonstrated in terms of the author's stance on academic controversy; his institutional affiliation ("major" vs. "minor" university); his nationality. These biases are small but statistically significant; it should worry us that the worst offenders were referees from "major" (i. e. large and well-funded) scientific departments.

The writer recently took a close look at the distribution, in geographical terms, of the subject matter of papers cited in the abstract service British Geological Literature over the period 1972-1976 (Brown, unpublished). Interesting patterns emerged. The Geophysical Journal of the Royal Astronomical Society has a remarkable interest in land and sea areas to the west of Britain. The Journal of the Geological Society showed a strong preference for the Northern Highlands of Scotland; a pattern quite different from that of the Proceedings of the Geologists' Association (an organization whose geographical focal point is the same meeting room in Burlington House). Of course one has to allow for historical bias in the development of individual societies, and for the greater geological interest of certain parts of the country. But the poverty of North Sea information in all but the trade journals and the pages of Nature was a really remarkable feature of the survey, only modified by very recent publications. It may be that these biases detract from the general utility and even the economic viability, of the journals concerned.

Tougher standards of refereeing and conscious selection of papers might do much to improve the quality of geological information in Britain. But who is to do it? The wealth of this country's geoscience documentation rests in the freedom of every organization to initiate new activity; the result is that no single body has the authority to set targets, goals or aims. Moreover the arrangements for information handling and control are piecemeal, whether one looks at libraries, abstracting services, serial publishing, or even borehole records.

Since 1972, when the Report of the Commission of Enquiry into the Research Councils headed by Lord Rothschild was published ("Framework for Government Research and Development" 1972) the Institute of Geological Sciences (IGS), the country's official body for geological investiga-

26

tions, has as a constituent part of the Natural Environment Research Council (NERC) been obliged to justify most of its activities in terms of their utility to specified clients or customers, either within government, overseas bodies or commercial enterprises. The "customer-contractor" principle has put increasing strain on the Institute's traditional role as the Geological Survey of Great Britain - mapping, recording and publishing the basic information about British geology. Particularly affected has been the programme of publication. Much IGS data is available to the public on limited access, but the overall effect of Rothschild has been to make the Institute more like a large commercial organization in its workings than a public service body, especially as concerns compilation of data of nation-wide scope. In many instances, this role is now being carried out by private organizations: Bibliographic Press Ltd compiles and publishes British Geological Literature from information provided by Brown's; the Committee of Heads of University Geology Departments compiles information on research projects at universities; geological thesis information has been brought together by Herrington Associates on a computer file at Sunderland Polytechnic and has been published as The Herrington List (Hodgson & Laming 1976); data on geological collection localities is being indexed by the Geological Curators' Group; and so on. Some of these compilations are partly duplicated by NERC (for instance a computer file of NERC-funded research projects is maintained); but these generally tend to be restricted to library holdings of a single NERC institute, or research funded by NERC - excluding other, nationally relevant, material.

The British National Committee for Geology of the Royal Society plays an important role in promoting international scientific co-operation through participation in international programmes and exchanges with foreign academies. Through the Council of the Society, it makes recommendations to government on aspects of geological science policy in the UK. But it is not in itself a document-originating or compilatory body, being formed of eminent scientists who have many other time-consuming activities. It could never therefore be the executive of a nationally co-ordinated policy for geological information: it might offer itself as the legislature for such a policy: but it is arguable that it does not by any means represent the user community for geological information, a community which extends far into other scientific and technical disciplines, and (no less important) into non-scientific and non-academic walks of life too.

A linking role among almost all of those who actively produce geological documentation, is played by the Geological Information Group of the Geological Society of London (GIG). The GIG has undertaken several important projects of collation and compilation, including a list of titles of early geological books, sources of information on Southeast Asian geology, and the Geological Directory of the British Isles (Diment 1978) mentioned earlier. But GIG is also a part-time association of busily employed people, and while its co-ordinating role is evident (if only from this well-attended meeting), it has no financial or manpower resources on which to draw in undertaking any comprehensive programmes of research; this seriously limits its effectiveness in such praiseworthy objectives as, for instance, the revision of UDC schedules for geology (a visionary project with which the author has been somewhat associated).

A newcomer to the scene in British geology, and potentially a valuable ally for those seeking improvements in the professionalism of geological documentation, is the Institution of Geologists (IG), incorporated in August 1977 and formally inaugurated at a national meeting in February 1978. The Institution is the result of extensive debate and consultation following the Report of a Working Group of the Geological Society on Professional Recognition (Geological Society 1974); it has been set up as the linking organization for those professionally engaged in geology in Britain and overseas, and as such it is concerned with the roles and activities of geologists as distinct from the study or promotion of the science of geology. As the "voice for geology", IG may be expected to play an increasingly important role in matters as diverse as the activities of public bodies; presentation of reports; and access to, and preservation of, geologically interesting localities. With its broad base in industry and the administrative services, IG has potentially the status required to press Government into funding projects and surveys of national scope; and if the GIG were to become more closely linked to the Institution, it could find itself in a powerful position to initiate projects of data collection and documentation of national importance - but it will only happen if IG fulfils its growth target and becomes the truly national professional geological body which it has set out to be.

When we turn to the organization of geoscience documentation itself, we again find great diversities. London has possibly the best concentration of geological libraries and collections of any single city in the world, many of them established for over 200 years (Hardy 1974, Robinson 1976). These include major national collections for science, patents, natural history and general reference; society and institutional libraries concerned with geology, geography, astronomy, mining and metallurgy, petroleum, fuel and engineering; university and college libraries (the Library Resources Co-ordinating Committee of London University recently compiled a list of all college and university serials holdings filling 33 microfiche); and Government Ministries and Departments. Providing a partial umbrella of co-ordination in such matters as inter-library loans is the British Library system (BL), which also performs important services in recording and documenting conference proceedings, translations, monographs and report literature (none of them confined to the geological sciences of course). The Science Reference Library provides, within the BL system, a service of access to major on-line abstracting and indexing services. Geology now has four main services, of which Geo-Archive, produced by Geosystems in London, is British. For technical reasons all four systems are not equally accessible.

Abstracting services cover a wide variety of subject fields in and related to geology. Already mentioned is the national regional bibliography which the writer's organization helped to revive in 1972; this is prepared without mechanical aids apart from a typewriter, by simple human effort, a feature which is held to be reflected in the quality of the abstracts (there are other systems which show evidence of mechanicality both in their input and their output). Transfer to a computer system may not take place before the advent of desk-top computers as standard office equipment. According to a paper by G Lea at this meeting, this stage is not far ahead. Abstract services are produced in Britain for mineralogy, rock mechanics, mining and metallurgy, petroleum, metals, ceramics, clays and coal (two series). The well-known GeoAbstracts produced in Norwich by Professor K. Clayton of the University of East Anglia has recently been extended by the revival of Geophysical Abstracts in a new format (reported in Geoscience Information Society Newsletter No 47, August 1977, p2).

News services are another important source of information, and in Britain we have several. Among the most useful are the Commonwealth Geological Liaison Office Newsletter, Geology Teaching (Journal of the Association of Teachers of Geology), Earth & Life Science Editing (Editerra/ELSE Newsletter) and the Newsletter of the Geological Curators' Group. British Geologist (Institution of Geologists) is particularly useful for hints about government enquiries in progress. Mining Journal, despite its title, is a weekly news magazine covering world wide developments in mining, and a regular source of information not only on mineral exploration and discoveries but on the background geological research and services. Also of value for their news content are the Newsletter of the Geological Society and the Circular of the Geologists' Association, both of which include items of news outside their own programme notes. Brown's Geological Information Bulletin, which is privately circulated to actively participating subscribers, attempts to draw together information on a number of aspects of current geological activity: research in hand, exploration results, new publications, meetings - a genuine "current awareness" in other words. Like all such attempts it is never as successful as one hoped, but it has brought the publishers into contact with a widening circle of kindred spirits.

Serial publishing in Britain shows on the surface a split between learned society and commercial publishers; but the two operate in symbiosis, having climbed a difficult learning curve in attempting to overcome some of the technical and financial problems of modern scientific serial publishing. The topic is well summarized in a recent study (Harvey 1978) which brings out the recurrent patterns of fluctuation in cash and enthusiasm that affect invisible colleges when they make the hazardous mutation into academic publishers. The recent EEC Seminar in Luxembourg on the future of publishing by scientific and technical societies, suggested to some UK observers that Britain was already some way ahead of her continental colleagues in coping with the problems of rising cost, diminishing readership for individual papers, and an increasing volume of scientific outpourings. Of course we have the advantage that we think and write in English - the international language of science. But we also have quite a long tradition of meetings and ad hoc organizations set up specifically to confront the practical problems of serial publishing - in which editors and publishers of geology have been active participants; one remembers particularly the meetings convened since 1948 by the Royal Society, as well as the variety of more recent conferences recorded in the news pages of Earth & Life Science Editing. This is not an excuse for smug complacency on our part, but may encourage those who feel inclined to throw up their hands in despair, to the effect that, on the contrary, where there's a will, there is also a way out.

Finally a data compilation topic, one close to the writer's own interests, which points up some of the difficulties we still face in Britain through lack of centralized thinking. Deep boreholes in and around the UK are by statute reported to the Institute of Geological Sciences, which keeps records and cores as a national archive. Since the beginning of offshore exploration for hydrocarbons more than one thousand wells have been drilled, and within the last two years some 300 of these as well as 28 on-shore wells have been published in the form of microfiche record sets, by the Department of Energy ("UK Continental Shelf Well Records" 1976-). Much of the material supplied to IGS is commercially confidential and therefore understandably inaccessible. The boreholes put down by IGS itself are however summarized in an annual report (IGS 1974, 1975, 1976) and later more fully published in separate bulletins (e.g. Poole 1978). Older material held by IGS is however published in a bewildering variety of reports, bulletins, memoirs and summaries of progress, not always written-up by IGS staff, and certainly not to a single standard as regards format, descriptive terms, nomenclature or degree of detail. Moreover there is no national data archive in the sense of a computer file which could be searched by correlative parameters in order to extract isopach, depth or facies information. Even if the existing, voluminous, material could be re-compiled to the same format as the Department of Energy microfiche records, we would at least have a single form of presentation from which geological interpretations could more easily be drawn. So far there is no sign that this haphazard system will change; in the meantime, a valuable resource lies relatively untouched. Perhaps renewed interest in onshore petroleum resulting from recent British Gas discoveries in Dorset, may trigger a new look at the existing record material and give weight to the desire for its general public availability.

Even between different departments of Government, therefore, failures of co-ordination are evidently not impossible. But, overall, there

are some solid advantages to be seen arising from the diversity of approach
in geological documentation activity in Britain. Firstly, there is room for
unconventional approaches to find expression; Brown's Service, combining
an exchange of information with a commercial service of library work and
bookselling, provides an example. Experiments are possible on a restricted
scale, which may point the way for future development by bigger organizations: for example, the Mineralogical Society's trial of "synopsis-cum-
miniprint" in Mineralogical Magazine (described in Earth & Life Science
Editing 6: 16). Collation and co-ordination is a necessity in circumstances
of diversified activity; the Geological Information Group, the Geological
Curators' Group, the Herrington List of theses, even the recently-started
listing of Books in the Earth Sciences (1977-)   -   all are attempts to come
to grips with aspects of this problem. And this collation may be the gateway
to closer co-operation and integration (though examples of the last are as
yet hard to find). But even if the integration of systems still seems a pipe
dream, Britain has not lagged in preparing the means for this to take place
(Jeffery, this volume). In the end, economics and utility may well enforce
the coming-together of diverse systems, both at the national and international levels; and the means will, in some instances, be provided by the
ingenuity of computer engineers.

Gradually, therefore, a pattern is taking shape in our geological
documentation activities. The diversity is being brought under control,
either by economic pressure or by the voluntary co-operative and compilatory efforts of individual bodies. Awareness of the problems is considerable, and the major bottleneck is organizational: we lack a central
body powerful enough to initiate major changes. But if the means can be
found to restore and enhance the sense of purpose and public spirit in our
national institutions through proper funding and technical support, Britain
will have a major contribution to make in future geological documentation.
The resources are there waiting to be tapped.

REFERENCES

Books in the Earth Sciences. 1977-, quarterly. Worthing : Bibliographic
    Press Ltd, Ferring, Worthing, Sussex BN12 5NH.
British Geological Literature. 1972-, quarterly. Worthing : Bibliographic
    Press Ltd.
DIMENT, J. 1978. Geological Directory of the British Isles. London :
    Geological Society, 109pp.
Framework for Government Research and Development. Cmnd 5046. 1972.
    London : HMSO.
Geological Society. 1974. Report of the Working Party on Professional
    Recognition. London : The Society, 28pp.
GORDON, M. 1978. Maintaining quality: refereeing. Earth & Life Science
    Editing 6 : 12-14.
HARDY, J. E. 1974. Libraries for the geologist in and around London. 3rd
    edition. London : Imperial College Lyon Playfair Library, 20pp.
HARVEY, A. P. 1978. A history of geological serial publishing in the
    United Kingdom. Earth & Life Science Editing 6 : 7-12.
HODGSON, A. V. & LAMING, D. J. C. Compilers. 1976. The Herrington
    List: Titles of research theses, 1960-75, on the geology of the
    British Isles and Offshore Areas, with classified subdiscipline/
    regional list. 2nd edition. Worthing : Bibliographic Press Ltd., 128pp.
IGS. 1974. IGS Boreholes 1973. Reports of the Institute of Geological
    Sciences 74/7. London : HMSO, v+23pp.
IGS. 1975. IGS Boreholes 1974. Reports of the Institute of Geological
    Sciences 75/7. London : HMSO, iii+26pp.
IGS. 1976. IGS Boreholes 1975. Reports of the Institute of Geological
    Sciences 76/10. London : HMSO, iii+47pp.
POOLE, E.G. 1978. Stratigraphy of the Steeple Aston Borehole. Oxford-
    shire. Bulletin of the Geological Survey of Great Britain 57. London :
    HMSO, iv+85pp.
ROBINSON, E. 1976. A guide to geology libraries in London. London :
    University College London Geology Department, Gower Street,
    London WC1E 6BT, 20pp.
UK Continental Shelf Well Records. 1976-. London : HMSO for Department
    of Energy, microfiches.

THE CURRENT AUSTRALASIAN EARTH SCIENCES

INFORMATION SCENE

DESMOND A. TELLIS

Australian Mineral Foundation, Glenside, South Australia, 5065

Summary: The national and State geological surveys have the largest collections of earth science information on the Australasian area. The paper
gives a brief review of new developments in information in the State
surveys and the national agencies in Australia. The present status of the
Australian Earth Sciences Information System (AESIS) and the Australian
Thesaurus of Earth Sciences and Related Terms is indicated. An outline of
the nature of information provided by the Papua New Guinea and New Zealand
geological surveys is also given.

The largest collections of earth science information on the Australasian
area are with the State and national geological surveys. Access to and
dissemination of information in these collections in Papua New Guinea and
in New Zealand are controlled by the national surveys. In Australia this
is done by each of the 6 States separately and by the Bureau of Mineral
Resources, Geology and Geophysics, in respect of its own information
resources.

AUSTRALIA

Conventional bibliographic coverage of most published earth science
monographs in Australia is provided by the Australian National Bibliography,
and material in the major earth sciences journals is covered by the Australian Science Index. However, much of the information in the earth sciences
in Australia remains in unpublished reports and whilst this material is
listed periodically, coverage and periodicity of announcement varies and
retrospective searching is difficult.

In an attempt to improve this fragmented situation the Australian
Earth Sciences Information System (AESIS) was developed in 1976 as a
national system in cooperation with the Bureau of Mineral Resources Geology
and Geophysics (BMR), the State geological surveys, the Commonwealth Scientific and Industrial Research Organisation (CSIRO), the National Library
of Australia, the Australian Geoscience Information Association, many companies and the Australian Mineral Foundation which initiated the scheme
and currently coordinates the system. AESIS provides bibliographic coverage of both published and unpublished material, the latter being mainly
open-file reports, theses and company technical reports.

An overview of Australian earth science information covering major
collections and systems, and including details of AESIS and the Australian

34

Thesaurus of Earth Sciences and Related Terms, has been given by Parkin
and Tellis (1977).

Activity since then has centred around greater use of the computer for
storage and retrieval of bibliographic and numerical data and increasing
recourse to microforms for easier storage and access to source material.

At BMR the Reference Mineral Collection Index now covers 5500 speci-
mens. The Index file is an INFOL system, as is also the Georgina Basin
Project file which contains over 3300 sets of field observations, generally
input in free field format. A comprehensive Palaeontology Index to BMR's
palaeontology collection is being developed on an HP2100 computer system
using the IMAGE data base software. Principal facts of over 200,000 gravity stations are now on magnetic tape. The earthquake data file contains
details of all earthquakes that have occurred in the region 0° to 90°S and
75°E to 165°E since 1897. The Bureau has commenced publishing its reports
and some appendices to Bulletins on microfiche.

In New South Wales the Water Resources Commission has a computer
based recording system in operation for 50,000 wells and bores using a
'Model-8' operating system on a Honeywell 8200. In the context of water
data, the Australian Water Resources Council (1976) has published standards for interchange of water resources data on computer media. A project
has been initiated by the N.S.W. Geological Survey for a computer based
system (Coalbor) for coal exploration data from the Muswellbrook area in
N.S.W.

The Geological Survey of Queensland is using the computer increasingly for scientific programming, computer graphics, numerical analysis
and management information systems. Activity in the information field
includes analysis of data from laboratory tests in geochemistry, engineering geology and hydrogeology; graphical applications associated with the
display of field data; modelling of groundwater flow and geophysical
problems; creation of data bases such as those operating for drainage
geochemical data and hydrogeological data. The Survey has a Tektronix
4010 VDU and a GE Terminet 1200 connected to a PDP-10 at the University of
Queensland. One of these terminals will soon be connected to the State
Government Univac 1100/40 dual processor.

The South Australian Department of Mines and Energy continues to
extend the use of computing systems to storage and control of its information sources: the bibliography index now has about 4000 entries; the
groundwater indexes consist of a Bore General File, Observation Bore File
and a Water Quality File, recording some 120,000 bores. Summary details
on each bore are included in the Bore General File with 25,000 entries on
magnetic tape so far - access is mainly via COM microfiche. The Observa-

tion Bore File can be used to present output as contoured plans of water elevation. The Core Library File has about 11,000 entries; Rock sample File some 10,000 entries; and there is a series of Geophysics Data Files, gravity data from which is used to produce Bouguer gravity contour plans. The Department has a CDC VDU and Tektronix 4051 Graphic terminal linked to the State Government CDC Cyber 73.

In Victoria the Department of Minerals and Energy is implementing a computer based file in conjunction with the State Rivers and Water Supply Commission to include 35,000 bores and 18,000 chemical assays. The system uses a Burroughs 7700 computer with a DMS 11 package. Planning is in the preliminary stages for the creation of data bases for bore information for engineering geology purposes, a fossil locations file and a mineral/rock locations file.

Access to information in Western Australia has been considerably increased by the recent introduction of reports in micro-form and the commencement in January 1978 of an active open-file system covering some petroleum, coal and other mineral exploration reports on 35mm film, currently representing about 60,000 frames. Also on open-file at present are 504 hard copy petroleum exploration reports. Unpublished and restricted information comprises about 7100 titles along with data on some 51,000 water bores.

The CSIRO and the National Library provide access to international data bases in the earth sciences, mainly of USA and UK origin. Internally, the CSIRO Division of Applied Geomechanics has created the GM Index - a machine readable data base - which covers the Division's publications, and the Division of Mineral Chemistry has developed CSIRO-THERMODATA a metallurgical thermodynamic data base. CSIRO's Central Information Service plans to have numerical data bases in crystallography, mass spectrometry and powder diffraction operating during 1978. Work on geochemical data and structural data files also continues within CSIRO as well as activity aimed at making computer-enhanced Landsat imagery more readily available in Australia. Also in the area of remote sensing the Remote Sensing Association of Australia is compiling a bibliography of Australian remote sensing literature, the persons concerned with this project are mainly from the BMR, CSIRO Division of Mineral Physics and from Technical Field Surveys Pty Ltd in Sydney.

Input to AESIS is proceeding at about 2500 entries a year. The BMR, all the State surveys, and the Papua New Guinea Geological Survey are now providing input to AESIS, besides many other organisations and companies. The CSIRO, in addition to providing the computing support for AESIS and progressive refinement of the programs in use, also assists in monitoring

material for input as does the National Library of Australia. Indexing
and retrieval is being studied and a seminar for the evaluation of AESIS
is scheduled for 5th and 6th October, 1978.

Australia is keen to know more about the characteristics and analysis
of earth sciences information to aid indexing, storage and retrieval, and
would be glad to participate in any area of cooperative study that may be
identified at this conference.

An up-dated print-out of a revised version of the Australian Thesaurus
of Earth Sciences and Related Terms is now being reviewed to produce a
first revised edition early in 1979. A COM version will also be available.

PAPUA NEW GUINEA

The main information resource for Papua New Guinea is the Geological
Survey of Papua New Guinea. The only other significant resource is the
University of Papua New Guinea library.

Access to the Geological Survey material is through card catalogues.
In addition, the following files are maintained:

1.   Papua New Guinea stratigraphic nomenclature register

2.   Gravity data

3.   Volcanic activity data

4.   Minerals index

5.   Well hole data (oil and water).

Regional seismic and magnetic data are held by the Bureau of Mineral
Resources, Geology and Geophysics, Canberra.

Geological Survey of Papua New Guinea publications include Memoirs,
1:250 000 geological map series and explanatory notes, 1:100 000 geological
map series, Notes on Investigations (1965-1973), and Reports (from 1973).
Unpublished material including reports from mining companies and professional opinions requested by various client organizations is held in the
data file collection. Earth science abstracts, Papua New Guinea have been
published for the period up to 1971 (Manser, 1974) and 1972-1973 (Manser
and Reynolds, 1976).

NEW ZEALAND

The New Zealand Geological Survey library is the oldest scientific
library in the country, being founded at the same time as the Survey itself in 1865. It currently holds some 30,000 volumes of monographs and
serials, 20,000 reprints, 3,000 geological maps, and receives about 1,100
serial titles annually. The library maintains a card index by author of
all material relating to New Zealand geology.

Main information sources are bulletins, geological maps, published
papers especially in N.Z. Journal of Geology and Geophysics (published
by the D.S.I.R.)   a semi-formal Report Series and an informal group of

37

reports dealing with a number of specific areas of work.

The Geological Map of New Zealand at 1:250 000 was completed in 28 sheets from 1959 to 1968; an explanatory text is printed on each sheet. 20 modern map sheets with separate booklet, at a scale of 1:63 360 are available, but continued regional mapping is now changing to a scale of 1: 50 000. Other maps include a recently-started Late Quaternary Tectonic series, and a small number of 1:25 000 sheets of urban areas, mainly Auckland.

Formal geological and palaeontological bulletins have been published since 1906 and 1913 respectively. Prior to this, geological information appeared annually in Reports of Geological Exploration and in the Appendices to the Journal of the House of Representatives. As the N.Z. Geological Survey is a Division of the Department of Scientific and Industrial Research, other specific items are published in the Department's Bulletins and Information Series.

The N.Z. Geological Survey Report Series, described above as semi-formal, is distributed to a restricted number of libraries and is available on request. It covers a wide range of topics from economic minerals to bibliographies and local areal geology, and includes a series of reports giving information about metallic minerals, non-metallic minerals and potential geothermal areas. Apart from published bulletins, now somewhat dated, general data on New Zealand coalfields are available in unpublished reports. Unpublished file information on economic topics as well as open-file reports from mining companies are accessed by a manual punch-card index. All petroleum company reports are held by the Geological Survey and up to date lists of open-file items are available on request. The New Zealand Mines Department administers the Mining and Petroleum Acts and is responsible for information on licensing and for the issuing of licenses.

Informal reports, dealing with specialities such as palaeontology, engineering geology, earth deformation studies and economic minerals, are produced in small numbers. The Volcanological Record is produced annually to make available data in the fields of volcanology and geothermal research.

ACKNOWLEDGEMENT

The author is indebted to the Directors/Chief Geologists and other officers in the State Surveys, the BMR, the Geological Surveys of Papua New Guinea and of New Zealand, and the Information Service of the CSIRO for so readily making available material for this paper.

REFERENCES

Australian Water Resources Council, 1976. Standards for interchange of water resources data on computer media. AWRC Hydrological Series No. 10, 92 pp.

Geoscience Information Seminar, 10-12 March 1975. Proceedings. Adelaide,
    Australian Mineral Foundation; 494 pp.

Manser, W. 1974. Earth science abstracts, Papua New Guinea, to 1971.
    Australia. Bureau of Mineral Resources, Geology and Geophysics.
    Bulletin 143(PNG 6), 445 pp.

Manser, W. and Reynolds, N.M. 1976. Earth science abstracts 1972-73.
    Papua New Guinea.Geological Survey. Memoir 4, 172 pp.

Parkin, L.W. and Tellis, D.A. 1977. Australian Earth Sciences Information
    System. Proceedings of the Australasian Institute of Mining and
    Metallurgy 262:7-23.

GEOSCIENCE DOCUMENTATION IN INDIA

S. B. GHOSH

Geological Survey of India

29, Jawaharlal Nehru Road,

Calcutta - 700016, India.

Summary: The paper deals with the history of geoscience research in India
from the days of Kautilya, Varahamihira to the present decade and the
involvement of different organisations in the country in its furtherance.
The present scene of the geoscience activities in the country is also
reviewed. The study of the history of documentation work reveals that
the documentation activity in India in the field of geoscience was
initiated during 3rd quarter of the 19th Century. The present state of
affairs in the documentation and information activities in post-indepen-
dence period and the different services rendered by the GSI at the
national level are discussed. Further role, the GSI has to play in the
organisation of a national information system in earth sciences under the
NISSAT plan of Govt. of India is mentioned.

0   Introduction

The evidence of geoscience study in India is recorded as far back
as centuries B.C.   The contributions of Kautilya (roughly in the 300 B.C.)
and Varahamihira (6th Century A.D.) etc. are still considered to be landmarks on the subject.   In fact, some of the present investigations are
undertaken on the basis of the revealation made by such earlier pioneers.
Varahamihira is sometimes considered as the 'first Indian earth scientist'
(Murthy, 1976).

I   Early Works

Subsequently, the Europeans travelling in the country made some contributions in this field.   The arrival of the British East India Company
in the country also resulted in the generation of geological information.
The first geological map in the country was published in 1814 (Heyne,
1814).   Though the first official appointment was made to survey the various parts of the country in 1817, with Mr. Laidlaw as the Mineralogist to
the Survey of Kumaon, the fulfledged regular geoscience activity could not
be started until 1851, which is considered to be the establishment of the
Geological Survey of India.   Dr. H.W. Voysey, sometimes regarded as father
of Indian geology, Capt. F. Dangerfield, Capt. Herbert, are some of the
geologists who made significant contributions in the field, prior to 1851
(Fermore, 1976).

Consequent upon the establishment of the Survey in 1851 with Dr. T. Oldham as the Director (though official designation was not Director), the Department started expanding and the tempo of the geoscience activity - both in research and documentation activity began. The first issue of Memoir was published in 1856, the Palaeontologica Indica, in 1861, Records in 1868. The first official geological map of the country was brought out in 1877.

II Societies/Associations - Their Role

The real tempo of scientific activity started with the formation of the Asiatic Society (of Bengal) in 1784. In 1808, two special committees, one of which was for the Natural History, Philosophy, Medicine were formed. In 1828, until when geological research was taken care of by this Committee, a Committee to promote geological researches was constituted. In 1847, standing committees of different subjects, one among them being 'Geology and Mineralogy' were formed. The society started a number of periodicals such as Asiatic Researches (1788), Transactions (1799), Journal (1832). The results of geological research were used to be published in these periodicals, the first contribution being of Leut. R.H. Coolbroke published in fourth volume of Asiatic Researches, in 1795.

Since 1851, with the establishment of major organisation in the country, the activities expanded. With the increased specialisation a number of societies were established. The first being the Mining and Geological Institute of India formed in 1905, the present name of which is Geological, Mining and Metallurgical Institute of India. Since then a number of societies have been formed such as Geological Mining and Metallurgical Society of India (1924), Palaeontological Society of India (1950), Geological Society of India (1958), Indian Academy of Geoscience (1958), Mineralogical Society of India (1959), Indian Society of Earthquake Technology (1962), Indian Geophysical Union (1963), Indian Association of Geohydrologists (1964), Geochemical Society of India (1965), Indian Society of Engineering Geologists (1965), Indian Geologists Association (1968), Indian Society of Photointerpretation (1969), Geological Institute of Presidency College (1905), Mysore Geologists Association (1950) etc. etc. All of them have their own publications.

III Geoscience Documentation

1 Early attempts

The first documentation work in the country was started in the field of geological sciences and done by a geologist in 19th century. Two bibliographies – Thermal Springs of India by T. Oldham (ed. by R.D. Oldham) and A Catalogue of Indian Earthquakes up to 1860 comp. by T. Oldham – were published in 1883.

The next worth mentioning attempt was made by R.D. Oldham when a comprehensive bibliography on Indian geology was published in 1888. The other attempts in the field of geological documentation are

1) A list and index of papers on Himalayan geology and microscopic petrology (1887).

2) Index to genera and species described in Palaeontologica Indica up to the year 1871 (1892).

3) Bibliography of barren island Narcendan (1895).

2 Twentieth Century

i) Pre-Independence Period

The most remarkable work in the field of geoscience documentation in India was done by T.H.D. La Touche in his monumental work Bibliography of Indian Geology (1917), in four parts. He also compiled a list of references on the Geology of Burma region, Index to Records (vols 1-65) of GSI (1932) and Memoirs (vols 1-54), published in 1936. He used the method of Keyword indexing in these indexes.

ii) Present Scene

Due to the increased demand of the natural resources and the necessity of their full exploitation for the development of the country, the need for organised information services was realised and the Government took major interest in the documentation service in the country.

The Indian National Scientific Documentation Centre (INSDOC), the national organisation in the field of documentation service published Insdoc List – Current Scientific Literature, from 1954 to 1966 and Bibliography of Scientific Publications of South and South East Asia from 1955 to 1964. The Indian Science Abstracts,

42

a monthly, is published by the Centre since 1955. All of them covered geoscience literature, but the coverage is not extensive.

The Geological Survey of India, being the only national organisation in this field till independence, carried out the major documentation activity. Since independence, a number of agencies in different branches of earth sciences have been established, as a result, a number of information services are being rendered by such agencies on different subjects.

A) Geological Survey of India

.1 The Survey which was started with 1 geologist, 1 writer and 1 peon has now a strength of 18,000 serving all over the country. It fulfils its objective through a well organised network of regional and circle offices spread all over the country. All of them are equipped with well organised libraries. The Survey during its long history of existence for more than 125 years made notable contribution in all branches of earth science, resulting in the generation of large volume of geological data and information. For utilisation of such information and to make use of the experience of others, it has arrangement to obtain world geological literature through subscription and exchange, with almost all earth science agencies throughout the world. The documentation service has been planned keeping in view the need of the geoscientists working in the country and is based primarily on the resources of the Central Library at Calcutta, with an integral network of regional and circle office libraries. The Central Library caters to the need of the scientists not only of the department but also of other institutions, with a resource of 3 lacs volumes of books, bound periodicals and other documents, such as unpublished reports, reprints, pamphlets, etc., with an annual addition of nearly 5000 volumes. It receives about 1500 serial titles from all over the world. Its services are rendered throughout the country through inter-library loan, reference and wide range of documentation and reprographic services.

.2 Central Information File : The library maintains a Central Information file of literature on Indian geology published throughout the world for easy retrieval and search service.

.3 Current Awareness Service : The users awareness services provided at the national level include

a) Library Bulletin, Ser A (Monthly), containing

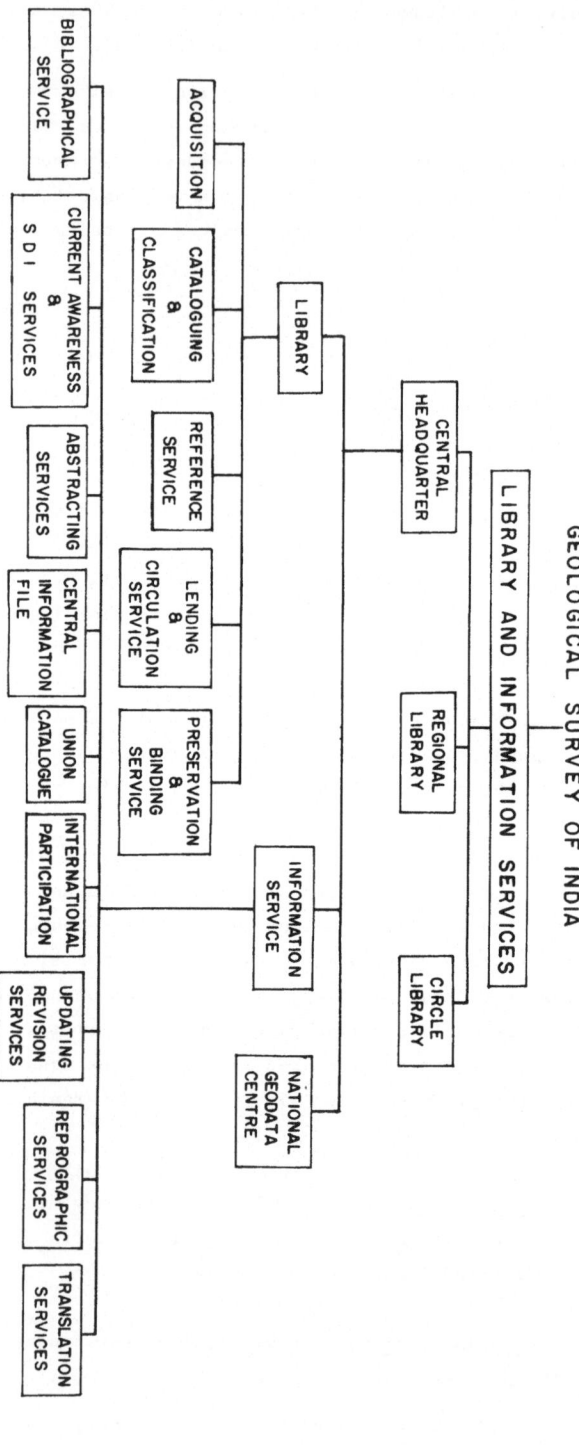

GEOLOGICAL SURVEY OF INDIA

LIBRARY AND INFORMATION SERVICES

PRESENT STATUS OF LIBRARY AND INFORMATION SERVICES OF THE GEOLOGICAL SURVEY OF INDIA

CENTRAL HEADQUARTER

REGIONAL LIBRARY

CIRCLE LIBRARY

LIBRARY

INFORMATION SERVICE

NATIONAL GEODATA CENTRE

ACQUISITION

CATALOGUING & CLASSIFICATION

REFERENCE SERVICE

LENDING & CIRCULATION SERVICE

PRESERVATION & BINDING SERVICE

BIBLIOGRAPHICAL SERVICE

CURRENT AWARENESS & S D I SERVICES

ABSTRACTING SERVICES

CENTRAL INFORMATION FILE

UNION CATALOGUE

INTERNATIONAL PARTICIPATION

UPDATING REVISION SERVICES

REPROGRAPHIC SERVICES

TRANSLATION SERVICES

current acquisition (macro documents).

b) Library Bulletin, Ser B (Fortnightly) containing
current index to Indian geological literature.

c) Earth Science Abstracts (Quarterly), containing
abstracts of selected earth science literature
relevant to the work of Indian geoscientists.

d) Indian Geoscience Abstracts (Annual).

SDI Service : The Central library and also, some of the
regional libraries provide project oriented and also, personalised SDI
service. For this purpose users'/project profile is maintained. The library is actively considering the plan to computerise the service in the
near future.

.4 Literature Search & Bibliographical Service : The literature search, both current and retrospective, and bibliography compilation
services are provided by Central, regional and circle libraries. The
bibliographies are regularly published in the departmental quarterly
'Indian Minerals' and also as separate publications. Mention may be made
of 'Bibliography on Himalayan Geology' and 'Geology and Mineral Resources
of Rajasthan'. A program on compilation of bibliographies on the geology
and mineral resources of different states of the country, for which the
latter forms a part, has been undertaken. This project when completed
will serve as a reference tool for the literature on geology and mineral
resources of the entire country.

.5 Survey on Earth Science Information Resources : Keeping in
view the proposed national information system, it became necessary to
have information on the existing earth science information sources and
services. Mention may be made of the following:-

1) Union Catalogue of Serials in GSI Libraries complex,
covering the holding of regional, circle and central
libraries 9 (published).

2) Catalogue of gazetteers available in Central Library
(published).

3) Compilation work of 'Union Catalogue of Earth Science
Serials in India' is under progress.

4) Inventory on Earth Science Information Services in
the country is under compilation.

5) Inventory of the unpublished reports brought out by the department is under preparation.

.6 _Translation Service_ : Studies have revealed that more than 25% of earth science literature appears in non-English language. To make use of such information, regular translation services from Russian, German and French languages are provided by the department.

.7 _National Geodata Centre_ : For utilisation of existing geo-science data generated in the country, the National Geodata Centre has been established. The data will be stored in the machine readable form for reasy retrieval and dissemination.

.8 _Reprographic Services_ : In order to make available the copies of the documents to the geoscientists working in different parts of the country, the reprographic facilities are provided with the help of modern reproduction equipments such as Xerox, Remington Copier, Scanner, Reader-printer etc.

.9 _Updating early work_ : The earlier documentation work carried out by the Department are being updated. _Annotated Index of Minerals of Economic Value_ compiled by La Touche in 1918 has been revised and publish-ed in 3 parts under the title _Annotated Index of Indian Mineral Occur-rences_ (as in April 1960). Updating the other works are in progress.

B) _Survey of India_

The Geodetic Branch Library (established in 1862) of the Survey of India caters to the need of the officers and staff of the organisation, by regular Current Awareness, SDI and News Clippings services. Repro-graphic facilities on a miniature scale is also provided.

C) _Oil & Natural Gas Commission (ONGC)_

Prof. N.A. Ermanko Library, established in 1957 provides cur-rent awareness, SDI, bibliographical, abstracting, news clippings and translation services on petroleum geology and exploration for oil in the land and offshore areas, not only to the departmental officers but also to others.

D) National Geophysical Research Institute

Established in 1962, it carries out research on various
geophysical aspects. It caters to the need of the scientists by providing
current awareness, bibliographies, abstracts, progress report, reprography and translation services. Observatories Data are published and
circulated amongst the scientists.

E) National Institute of Oceanography

The Institute established in 1966 provides the various
documentation services with the resources of 10,000 volumes of documents
and 250 titles of periodicals. Amongst the different functions, the
development of the Indian National Oceanographic Data Centre as a Central
Agency for oceanographic data, functioning as a national facility for the
exchange and dissemination of oceanographic data and information to the
user community and exchange of oceanographic data with the World Data
Centres A & B (Washington and Moscow) are included in the program of its
information and data division.

F) Indian Institute of Magnetism

Established in 1972 to carry out research in the field
of geomagnetism, it publishes Indian Magnetic Data.

G) Indian Bureau of Mines

Established in 1948, the organisation was set up for
handling problems of mineral industry and mining and collection and
maintenance of information and statistics relating to mineral trade etc.
Its Central library provides current awareness, bibliographical, abstracting, SDI services and brings out bimonthly documentation notes - Mines &
Minerals, publications added to central and regional libraries (quarterly)
and Reader's Profile Service. Data on mineral production, stocks,
imports, etc. are stored in machine readable form.

H) Central Mining Research Station

Established in 1955 to carry out research in mining
engineering, testing and certifying equipments for use in underground
mines etc., the library with a collection of 16,000 documents provides

fortnightly current awareness – Current Contents: Mining and allied subjects; a digest service – MINSCOPE – featuring all aspects of mining and
related areas of interest to the institute. Besides the above, it maintains data-base on energy, mines, mining methods, minerals and metals etc.
An alerting service 'Brought to Your Notice' is provided.

I) Birbal Sahani Institute of Palaeobotany

Established in 1946 it caters to the need of scientific
community in the field of fossil plants and their application to the problems of economic geology, coal and petroleum prospecting, through current
awareness, SDI, news clipping service, etc.

J) Other agencies

Besides the above, there are other organisations such as
Central Ground Water Board with a network of offices throughout the country, public sector organisations, such as Mineral Exploration Corporation,
National Mineral Development Corporation etc., State Govt. geology departments, etc. who are engaged in the specific branch of activity. All of
them have libraries and documentation services according to the need of
the scientists.

IV NISSAT and its impact on Earth Science Information System

A review of overall documentation activity in the country reveal that
though most of the activities have been initiated from government organisations, no definite organised efforts were made until recently. Realisation of the importance of organised information activities by the government and its effort for a planned information service opened a new era in
the overall information activity in the country. During the Fifth Plan
period the Govt. has launched a programme for building up a strong
national network of documentation and information services – National
Information System on Science and Technology (NISSAT), with a set of
objectives, to meet the information needs, of all – both present and
future. For this purpose a national Scientific and Technological Information policy has been formulated: "The NISSAT Network would be built up
on the existing infrastructure by integrating and co-ordinating the available services and activities and by seeking cooperation of individual
information centres. Weak links and gaps would be identified for strength-
ening them or for establishing new services". (Appukuttan, 1977). In

this network, sectoral system, regional system and other specialised services will be established. In the first phase, work for building up information systems for four sectors – leather, food, machine tools and drugs & pharmaceuticals has already been initiated. Amongst the others, this plan has identified the natural resources sector to be comprised of the subjects oil and natural gas, minerals, coal, ocean resources, water resources, geophysical and geochemical resources, etc., for which building up of information system will commence gradually.

With the increasing need of effective information services in the field of earth sciences, it is envisaged that a national information system – Indian Earth Science Information System (INESIS), comprising of a national centre, linked with regional and branch centres, will be possible in the near future.

The Geological Survey of India, the major organisation in this branch of science in general, will naturally play a vital role in the System.

Acknowledgement: Sri A.R. Chakrabortty, Librarian, Geological Survey of India,, suggested the topic and went through the manuscript. Sri S.K. Kapoor, Sr. Librarian, showed keen interest and continuously encouraged in preparing the paper.

My gratefulness is due to Sri K. Ranganathan, Director (Publications) and Sri V.S. Krishnaswamy, Director General, Geological Survey of India for kind permission to present this paper.

REFERENCES

APPUKUTTAN, N. 1977. National Information System in Science & Technology (NISSAT). (Country Report presented at 2nd UNISIST Meeting on Planning & Implementing of National Information Activities, Friedrichsdrof.) 27pp.

ASIATIC SOCIETY OF BENGAL. 1885. Centenary Review. Part 1. Calcutta : The Society.

CHAKRABORTTY, A.R. 1971. Early attempts of documentation of geological literature in India. Annals of Library Science & Documentation 18 : 132-40.

DEPARTMENT OF SCIENCE & TECHNOLOGY. 1977. Directory of data centres in India.

FERMOR, L.L. 1976. First twenty-five years of the Geological Survey of India. (GSI Miscellaneous Publication No. 39.)

HEYNE, B. 1814. Tracts, historical and statistical on India. London : Robert Baldwin etc., 462pp.

KAPOOR, S.K. & CHAKRABORTTY, A.R. 1977. Information system for earth sciences. In KAPOOR, S.K. & GHOSH, S.B. Editors. Planning of National Information Network. Calcutta : IASLIC.: D31 - D45.

MURTHY, S.R.N. 1976. Earth science studies in ancient India - an outline. Indian Minerals 30 : 29-42.

RANGANATHAN, S.R. 1963. Documentation and its facets. Bombay : Asia, 639pp.

HISTORY OF GEOSCIENCE INFORMATION IN INDIA

K. S. MURTY

University department of Geology,

Law College Compound,

Nagpur 440 001, India.

Summary: Ancient literature of India contains many references to fundamental concepts and facts of Geology, Mineralogy and Metallurgy. The
scientific achievements and mining activities of the medieval period were
described by Alberuni and Marco Polo. Modern geology and allied subjects
were introduced in the Indian Universities in the last quarter of the
19th Century while the Geological Survey of India, founded in 1851
brought out the first publication, a Memoir, in 1856. Growth of geological publications progressed and almost synchronised with the
establishment of more and more institutions and societies. The INSDOC is
the main documentation centre in the Country and publishes a Directory of
Indian Scientific Periodicals annually. Among the principal national
agencies which are involved in collection, storage and retrieval of geological information are the GSI, IBM, NGRI, ONGC, CMP&DI, and NIO. A
National Information System for Science & Technology (NISSAT) was
established under the Department of Science & Technology (DST) in 1977 to
integrate and coordinate the existing and future scientific and technical
information sources, systems and services into an organised and effective
network. It publishes a Quarterly news letter. Many of the State
Language Akademis are also publishing books on Geology, written
originally or translations from English.

A study of the ancient literature of India reveals that savants were

familiar with fundamental concepts and facts of Geology and Mineralogy,

and even Metallurgy. The origin of many scientific principles can be

traced in the Vedas (1500-2000 B.C.), while the excavations at Mohenjo-

daro (seven layers of buildings dating to 2750 B.C.-3250 B.C.) revealed

articles like jewellery made of gold, silver and copper, and glazed

pottery, and inscribed seals all of which indicate the high level of the

Indus Valley of Civilization. Words for melting and tempering were known

to Rgveda - AYCHATA (IX.1.2) and DHAM and SANADHAM (X.72.2; V.9.5).

The Sankhya-Patanjali system explains the principles of cosmic evolution

and the origin of the Universe. Kanada (600 B.C.), in his Vaiseshika

philosophy, explains the Paramanus as entities which cannot be divided

any further, are eternal and indestructible and make up everything in the

Universe. The Paramanus combine in pairs, triads, tetrads etc. Susruta

(600 B.C.) categorised the Earth substances which include gold and the

five Lohas. He also described the preparation of metallic salts.

Nagarjuna's Lohasastra (Metallurgy) (400 B.C.?) advanced the knowledge of

chemical compounds by his preparations of mercury. To Patanjali was

attributed a treatise on Metallurgy which deals with, among other things,
the extraction, purification and assaying of metals. The Rāmāyana (200
B.C.-200 A.D.) gives us abundant information on concepts of geological
and anthropological significance and on meteorological phenomena and
metals.

Chapter XII of Kautilya's Artha Śāstra is completely devoted to mining
operations, mine management and manufacture of metals. While Charaka
(200 B.C.?) classified animals in four different ways on the basis of
their habitat, mode of birth and other factors, Umāsvati (around 40 A.D.),
in his work Tattvārthadhigama, goes further to classify the animals also
on the basis of active sense development and on the basis of their
reproduction. Varāhamihira (499 A.D.-587 A.D.) was a versatile scientist
who dealt with astronomy, astrology, agriculture and influence of weather
in his works, of which Brihatsaṁhita is the most well known. He expressed that the earth was globular in shape, remained fixed while the planets
revolved around it, a view contradicted by Āryabhata. Some, including
Varāhamihira, hazarded the guess that the precious stones are rocks
metamorphosed by natural processes in the course of ages (Kōchit Bhuvah:
swabhāvat Vaichitryam Prāhurupalānam). Vāgbhatta (1300 A.D.) proposed,
in his Rasaratnasamucchaya, a division of the mineral kingdom into four
classes: eight Rasas, eight Uparasas, gems, and metals and their alloys.
Rasārnava, Rasaratnākara, Rasendra Choodāmani, Rasa Kalpa and Rasa Rāja
Lakshmi are some of the works that were written during those two hundred
years. Marco Polo describes how diamonds used to be mined during the
reign of Rudramamba (Kakatiya Dynasty) from Warangal up to 1296 A.D.

The dark period probably set in after the invasion of India by Mahmud
Ghazni. Centuries later, the work of early orient scholars like William
Jones, Pricep, Max Mueller, Wilson, and Ferguson and the early archae-
ological excavations in 1861 brought home to the educated Indians a vivid
picture of the past glory and greatness of ancient India.

Geological and related studies were introduced in India in the last quarter of the eighteenth century when the Universities were established in
Bombay, Calcutta and Madras. Sir Thomas Holland headed the separate
faculty in Geology in 1892 in the Presidency College, Calcutta. Today
nearly 7,000 Geoscientists are there in the country.

The origin and growth of the publication of scientific periodicals in
India almost synchronised with the establishment and development of

52

scientific institutions and societies. The Geological Survey of India, founded in 1851, was the first to bring out a publication, its first Memoir in 1856 to be followed by its Records and Palaeontological Indica (1861). Publication activity showed a corresponding increase as the tempo of scientific research increased. Many Universities started to publish their own periodicals. From 12 periodicals at the end of 1940, the number rose to 18 by 1950 and by 15 more by the end of 1960. The decade 1961-70 recorded a phenomenal growth in the number when as many as 27 periodicals were added. Many new Societies came into existence and some of the Government Undertakings also started publishing their own journals during that period. The present position is: 11 Monthly, 17 Quarterly, 5 Half-yearly and 18 Yearly periodicals while the rest are of differing frequency (Appendix III). Other General Periodicals, numbering about 12, also cover articles on Geological subjects.

Most of the University Departments of Geology publish journals/bulletins, but not very regularly. The Indian Science Congress Association brought out two Progress/Review Reports (1963 & 1972) while seven Commemoration Volumes, containing valuable articles by Indian and foreign geologists, were released on different occasions between 1961 and 1973 in honour of distinguished geologists. The Indian National Scientific Documentation Centre (INSDOC) publishes a monthly journal of Indian Current Science Abstracts. The Indian Documentation Service has brought out two issues of Indian Science Index (1975 & 1976) covering all sciences. The Geological Survey of India has been publishing the 'Earth Science Abstracts', selected from Indian and foreign journals since 1975. The Geological Society of India has been giving, since 1976, a list of research papers of major interest from Indian and foreign journals in its monthly publication, The Journal of the Geological Society of India. Three volumes corresponding to the years 1971, 1972, and 1973 respectively, of The Indian Geological Index have so far been published. The Indian geologist is now able to keep himself posted with the latest developments in the field of Geology much sooner than before through these publications and other Bibliographies (Appendix I). A few Glossaries in Indian languages have also been published to help in writing of books on Geology in regional languages, while a few Directories and Handbooks are also available for ready reference (Appendix II).

Apart from the 50 and odd books in English on the various branches of Geology, there are also about 70 text books and laboratory manuals (original and in translations) in Indian languages up to the degree level, majority of them being in Hindi and Telugu.

INSDOC is the modal agency in India for cooperating with the International Serials Data System set up under the sponsorship of UNISIST/UNESCO. It uses IBM 360 Model 44 Computer (PL/1 Programme) developed at INSDOC. The arrangement is according to the Universal Decimal Classification Scheme and the entries, under each subject, are arranged alphabetically. Its latest edition (1976) of the <u>Directory of Indian Scientific Periodicals</u> lists 1593 titles in all languages, some of them being multilingual too. The language periodicals number 260. Besides supplying photocopies or translations of articles required by laboratories or individuals, INSDOC also acts as a channel to let the world know of the scientific work done in this country. It has a branch in Bangalore. A one-year advanced training course in Documentation and Reprography is being given since 1964. More than 600 scientific libraries are located in important cities in India while about 83 Universities have developed their libraries under substantial support from the University Grants Commission and under the Wheat Loan Programme. The cost of the books, however, is going beyond the reach of the students, if not of the teachers!

Some of the principal National Agencies which are involved in collection, storage and retrieval of geological information include the Geological Survey of India (GSI), Indian Bureau of Mines (IBM), National Geophysical Research Institute (NGRI), Oil and Natural Gas Commission (ONGC), Coal Mines Planning & Design Institute (CMP&DI) and the National Institute of Oceanography (NIO).

The GSI has a Coding and Retrieval Section that acts as a Data Centre. It uses a flexible system of classification and coding and prepares punched card system, with designs for storage and retrieval of geological data. Recording of this data on magnetic tapes is proposed to be done after adequately testing the punched cards system for its efficiency, sometime after 1978. Files have been designed on samples, field data, palaeontology, geochronology, chemical analyses, systematic mapping and mineral inventory. The GSI has also prepared or compiled bibliographies on many topics relating to foreign or Indian geological matters. It has also started publishing in its <u>Indian Minerals</u> a feature under the title of Stratigraphic Lexicon, individual stratigraphic horizons that have importance in regard to occurrence, lithology, etc.

The IBM has a Reprography Unit and publishes most of its data in its own

monthly publications like __Mines and Minerals__. It also employs the
punched card system using an IBM machine, with storage being done in
Pune. One of its latest publications is titled 'Indian Mineral Industry
at a Glance', a multi-coloured graphic work.

The NGRI stores and retrieves data from punched cards.

The ONGC has an IBM 370-145 and has been working on a comprehensive
Computerised Petroleum Data Bank (COPED). A General Data Management
System (GEDAMS) is currently under development.

The National Institute of Oceanography, in cooperation with FAO/IOC Panel
of experts on the Aquatic Sciences and Fisheries Information System, is
trying to evolve an Information System in accordance with the UNISIST Concept. These organisations have been selected to maintain data banks on
minerals and ores so as to avoid duplication of work and to help arrive
at uniformity and standardisation of storage and retrieval systems.

A National Information System for Science & Technology (NISSAT) was
approved by the Government in May 1977 and it started functioning under
the Department of Science & Technology (DST) with a view to integrate and
coordinate the existing and future scientific and technical information
sources, systems and services into an organised and effective network.
Under this, a Mineral Data Bank is envisaged, details of which are being
chalked out. It is also proposed to have Sectoral Information Centres
among which figure the Natural Resources and Energy separately. NISSAT
publishes a Quarterly news letter.

A suggestion was made (Krishnamurthy, Ganeshan, Ramaswamy & Iyengar:
1976) for the establishment of a National Earth Sciences Data Centre
(NESDC) to act as a repository for all the Earth sciences data emanating
from GSI and other organisations in the country like the State Director-
ates of Geology and Mining, ONGC, IBM, AEC, Universities, and Public
sector undertakings like NMDC, CIL, MEC etc. It would be responsible for
the data collection, storage, retrieval and dissemination of data and
also act as a focus for exchange of data, information and ideas.

Within two years of the Seminar on Indian Programme for Space Research
and Applications held from August 7 to 12, 1972 at Ahmedabad under the
auspices of the Indian Space Research Organisation (ISRO), the Government
of India established the National Remote Sensing Agency (NRSA) at Hydera-
bad for making use of the modern technology of remote sensing for quick
and repetitive collection of data pertaining to natural resources of the
country. NRSA's equipment now consists of 3 aircraft, sensor systems such
as Bendix Multi-Spectral Scanner, Magnetometers and aerial cameras and a
computerized image analysis system (Bendix M-DAS). LANDSAT imageries and
CCTs are now being bought from the NASA USA. The remote sensed data is
used for getting inventories of agricultural crops, forests, water
resources and mineral resources, for land use planning, updating of maps
etc. The NRSA has plans to build up a data bank for storing of the
information collected through remote sensing and for retrieval when
required.

The Indian and American Governments exchanged diplomatic notes on January
3, 1978 during President Carter's visit, confirming that the USA would
programme its LANDSAT earth resources satellite to transmit data directly
to a ground receiving station that India would own and operate. This
satellite service is expected to provide India with comprehensive topographic and mineral information and timely data on the everchanging condition of agricultural, water and other natural resources.

There are plans to include a scanning radiometer in the second Indian
Scientific satellite to be put in orbit in 1978 from a Russian base.
Called the Satellite for Earth Observations (SEO), it will use appropriate
sensors to observe the earth in the visible, near infra red and microwave
frequency bands to obtain relevant scientific information in the areas of
meteorology, hydrology and oceanography using the satellite born sensors.

Studies have been made with the help of Apollo photographs and Landsat-1
imagery in selected parts of India in regard to applications in Geology,
drainage patterns and land use classification. Interpretations of multi-
hand imagery in photographic form and quantitative analyses were carried
out with the help of Computer-compatible tapes (CCT). The areas specifically studied include the northern Terai area of Uttar Pradesh, and the
Punjab. Geological maps were also made of areas in the southern parts of
Tamil Nadu and Andhra Pradesh States, showing various linear features and

some lithology. A detailed study was also made of the drainage pattern in the Saurashtra area using LANDSAT-1 imagery.

The NRSA has made use of satellite imagery to conduct an extensive survey of the upper Barak river watershed in the Surma valley of Assam. This would provide basic information for the integrated development of the Barak catchment area in the States of Manipur, Mizoram, Assam and Naga-land. SEO data is also expected to help government guard against cyclones and also flooding of the Ganges, the Jamuna and the Kosin basins by keeping a watch over the data from this 'Eye in the Sky'.

Landsat-1 imagery was also used in the study of the Pre-Cambrian hard rocks and desert terrain of Rajasthan, the Anorthosite body of Bolangir in Orissa and the Koyna area. ERTS-A imagery was used in studying the structure of the Vindhyans in Madhya Pradesh, Forest Resources Surveys in Central India, Kerala, Andhra Pradesh, Manipur etc. and in Morpho-Tectonic analysis of the area around the Jalasindhi Dam sites in M.P. and Maharashtra.

Translations

In accordance with the policy that education at least up to the degree level should be in the regional language/mother-tongue, several State Governments established Text Book Akademies to plan and produce text books in their respective languages. Nearly 70 such books have so far been published in Telugu, Tamil, Hindi, Oriya, Malayalam, Marathi and Bengali, after preparing the necessary Glossaries of terminology.

Some standard works have also been translated by these Akademies. 'General Geology' by O. Lange, M. Ivanova, and N. Lebedeva has been trans-lated into Tamil, Crumbein's 'Introduction to Geology', Platt and Challinor's 'Simple Geological Structures', G. W. Tyrrell's 'Principles of Petrology', and A. Johannsen's 'Descriptive Petrography of Igneous Rocks' into Hindi, and Lahee's 'Field Geology' into Telugu.

The progress has been remarkable in the case of Telugu and Hindi books, but slow with respect to other languages.

ABSTRACTS AND INDEXES

Title	Publisher	Year of commencement	Frequency
Earth Science Abstracts	GSI, Calcutta	1975	M
Indian Geological Index	Indian Geological Index, Delhi 32	1971	Y
GSI Library Bulletin, Sections A & B	GSI, Calcutta	1952	F
IBM Library Bulletin	Indian Bureau of Mines, Nagpur	1956	M
Index to Selective Articles	IBM, Nagpur 1	1960	M
Indian Current Science Abstracts	INSDOC, New Delhi	1965	M

BIBLIOGRAPHIES

Title	Publisher	Year of commencement	Frequency
Bibliography of Hydrology in India, 1936-52	Ministry of NR & SC, New Delhi	1955	–
Bibliography of Indian Geology, Vols. 1 & 2	Manager of Publications, New Delhi 1	1969 (Rev. Ed.)	–
Bibliography of Mysore Geology	Mysore Geologists' Assn., Bangalore	1955	–
List of Publications	GSI, Calcutta 26	1963	–

M: Monthly   F: Fortnightly   Y: Yearly

Title	Publisher	Year of Publication
Dictionary of Geology Ed: Sen Gupta, S.C. English-Hindi	Hindi Sahitya Sammelan, Allahabad	1953
Glossary of Geological Terms English-Marathi	Maharashtra Textbook Bd., Pune/Nagpur	1972
Glossary of Geological Terms English-Telugu	Telugu Akademy, Hyderabad 29	1971

DIRECTORIES/YEAR BOOKS

Annotated Index of Indian Mineral Occurrences, as in April 1960, Pts 1-4 Comp: Chatterji, P.K.	GSI, Calcutta 26	1962/64
Coal Directory of India	Coal Board, Calcutta	1969
Directory of Indian mines and metals Comp: Ghosh, P.K.	Mining, Geological and Metallurgical Institute of India, Calcutta	1952
Indian Mineral Year Book (English and Hindi)	IBM, Nagpur 1	1955 onwards

GENERAL

Directory of Indian Scientific Periodicals	INSDOC, New Delhi	1976
Indian Science Index	Indian Documentation Service, Gurgaon/New Delhi	1975 (Yearly)

APPENDIX III

List of current periodicals – not reproduced

REFERENCES

ALI, S.M. 1961. Geography in Ancient India. Symposium on the History of Sciences of India. National Institute of Sciences : 258-280.

BRAJENDRA NATH SEAL. 1958. The Positive Sciences of the Ancient Hindus. Delhi: Moti Lal Banarsi Dass, 295pp.

DEPARTMENT OF SCIENCE & TECHNOLOGY. 1977. National Information System for Science and Technology (NISSAT) - Country Report: 28pp.

DIPANKAR LAHIRI. 1970. Dynamic Geology in Pre-historic India and Vedic Literature. Indian Journal of History of Science 5:

INSDOC. 1976. Directory of Indian Scientific Periodicals. New Delhi: INSDOC.

KRISHNAMURTHY, GANESHAN, RAMASWAMY & IYENGAR. 1976. National Earth Sciences Data Centre, Collection, Storage, Coding and Retrieval of Data Dissemination of Information. National Earth Sciences Register, 7pp.

MARCO POLO. The Travels of Marco Polo. New York: Books Inc.: 287-289.

MIRA ROY. 1961. Scientific Information in the Ramayana. Proceedings of the Symposium on the History of Sciences of India. National Institute of Sciences : 58-66.

MURTY, A.T. 1974. Sources of Information for Geological Literature in India. Journal of the Institute of Geology. Vikram University, Ujjain : 97-104.

MURTY, K.S. 1977. Trends in Earth Science Education in India. In COORAY, P.G. Editor. Geoscience Education in Developing Countries, Report No. 5. St. John's: AGID.

NILAKANTA SASTRI, K.A. 1967. Cultural Contacts between Aryans and Dravidians. Bombay: Manaktalas.

PEARL, R.M. 1951. Guide to Geologic Literature. New York: McGraw-Hill Book Co. Inc., 239pp.

SASTRI, SHAMA R. Editor & translator. 1909. Kautilya's Arthasastra. Mysore : Book II, Chap. XII : 75-89.

UNESCO. 1972. National Science Policy & Organisation of Scientific Research in India. 27 : 19-22.

VIJAYENDRA RAMAKRISHNA SHASTRY. 1961. Science in the Vedas. Proceedings of the Symposium on the History of Sciences of India. National Institute of Sciences : 94-104.

GEOLOGICAL DOCUMENTATION IN BRAZIL

NIZA S. JARDIM

and

MARCIA R. MIGLIORATO

Instituto de Pesquisas Tecnológicas, S/A Divisão de Minas e Geologia,

Aplicado, Grupo de Documentacao Cientifica, 01000 São Paulo, Brazil

Summary: Geological documentation in Brazil remains decentralized and the responsibility of a number of university and government libraries. A major cumulative bibliography of Brazilian geology is nearing completion.

Although both geology and geological documentation have a long history in Brazil it was not until 1954 that the Instituto Brasileiro de Bibliografia e Documentação (now Instituto Brasileiro de Informaçcio Científica e Technologica) was founded to act as a specialized documentation centre, with a remit which included the compilation of both bibliographic and library science texts. This institute is also attempting to promote the exchange of information between the various research institutions and national and international documentation centres (Fonseca, 1973).

However, in 1951, before the foundation of the Institute, the central library of the University of São Paulo attempted to function as a documentation centre. This is an activity which it still continues, publishing a catalogue of serial holdings.

Geological documentation in Brazil, which remains decentralized, has been reviewed by Jardim (1973) and Santos and Jardim (1975). Major library directories have been compiled by Araujo (1969) and the Instituto Brasileiro de Bibliografia e Documentação. There are 44 libraries in the geosciences, of which 22 are in universities, 19 are in government institutes and 3 in the public domain.

The major libraries are:

1. Library of Departamento Nacional da Produção Mineral. The Departamento Nacional da Produção Mineral, a body of the Minis-

try of Mines and Energy, has offices in 9 regions of Brazil and
one library in each office. Its first library was in Rio de
Janeiro (1927). Until 1972 when it suffered a fire and was
totally destroyed it was the largest, most important and most
complete library of geoscience in Brazil. At that time it
possessed 187,000 volumes and 2,000 periodical publications.
Since 1974 periodical subscriptions have not been renewed. The
DNPM's library in Brasília, currently the main office's
location, has not centralized the services or coordinated the
various libraries.

2. Library of Instituto Geológico. Instituto Geológico is a body
    of the State Department of Agriculture of the State of São
    Paulo. It was created in 1886 under the name of "Comissão Geo-
    gráfica e Geológica". The foundation of its library dates back
    to the same year and it was entirely restructured in 1941. The
    library holds 70,000 books and pamphlets as well as 1,954
    periodical publications and 22,000 maps.
    It serves the Instituto's technicians as well as Universities
    and other entities. It publishes semi-monthly lists of new
    acquisitions and edits a monthly bulletin. It is the largest
    and most important geoscience library in the State of São Paulo
    and currently the most important in Brazil.

3. Library of the Instituto de Geociências of the University of
    São Paulo. Instituto de Geociências of the University of São
    Paulo-IGUSP, was one of the first schools of geoscience in
    Brazil being founded in 1957. Its library, however, is divided
    into several departments and only since 1970 has it been
    centralized. The stock consists of 15,618 books and pamphlets
    and 160 periodical publications. New stock is mostly added by
    exchange with other institutes.

4. Library of Companhia de Pesquisa de Recursos Minerais. The
    Companhia de Pesquisas de Recursos Minerais-CPRM, is a company
    linked to the Ministry of Mines and Energy, and was created in
    1970 with the aims of basic geological research and to serve
    individual and specific researches. It is divided into 9
    regional superintendencies, each of which has its own library.
    The main library is located in Rio de Janeiro and has 1,366
    books, 300 periodical publications and a technical file cabinet

with 700 reports of the Company itself.

Among the services rendered to users are indexes and summaries of articles received. These are also sent to all regional superintendencies.

5. Library of Instituto de Pesquisas Tecnológicas do Estado de São Paulo S.A. The Instituto de Pesquisas Tecnológicas (IPT) was founded in 1934 with the object of rendering assistance to the country's technology.

It is a company linked to the State Department of Culture, Sciences and Technology of the Government of the State of São Paulo.

Among its activities are, for example: technological support to large infrastructural works; technological assistance for the development of industrial processes; backing services to the development of machines; technological assistance in housing programs, and standardization. IPT's library is a section of the Divisão de Documentação Científica which also has a standards office and a Publication and Printing office.

It owns the most complete collection of books (80,000 volumes), 1,300 periodicals and many publications in the technological field.

It publishes a monthly accessions bulletin, using a KWOC index (listing authors, titles and key-words).

Dissemination services for scientific information include: Geodex Information Service; Pascal System, of the Centre National de la Recherche Scientifique; the DATRIX II of Xerox University of Microfilms; OUIPT - Biblio Index, of Union Nationale des Transports Publiques.

The library has developed a system of controlled decentralization of its collections, having created ten units to serve specific categories of users.

One of these units is the documentation group of the Divisão de Minas e Geologia Aplicada (DMGA), which has 6,000 books and pamphlets, 80 periodical titles, 150 congress proceedings; 2,000 slides; 2,000 maps, together with aerophotos and satellite images (LANDSAT).

Among its main services to users are the bibliographical catalogues, indexing and abstracting of articles in journals.

It maintains an exchange programme with almost all the geoscientific libraries of the country, and also provides material through inter-library loan.

One of the first organizations concerned with the interpretation of the geology of Brazil was the Comissão Geológica do Império do Brasil (Imperial Geological Commission of Brazil) which was founded in 1875 and only disbanded in 1977. In 1907 the Serviço Geológico e Mineralógico was established and, in 1933, this became part of the Departamento Nacional de Produção Mineral (of the Ministry of Agriculture). This department embraced the following divisions: Serviço Geológico e Mineralógico, Serviço de Fomento da Produção Mineral and the Serviço de Águas e Laboratório Central. The Serviço Geológico e Mineralógico was in fact created in 1907, its first bulletin being issued in 1920. This later became the Divisão de Geologia e Mineralogia.

In 1939 the Conselho Nacional do Petróleo was established to promote the search for oil and to formulate a coherent national policy. The Conselho Nacional de Pesquisas, a research agency, was founded in 1951 to provide a central incentive for research in geology. In 1955 it initiated a programme to search for atomic minerals and later became the Comissão Nacional de Energia Nuclear.

In 1960 a reorganization resulted in the creation of the Ministry of Mines and Energy which absorbed the Departamento Nacional de Produção Mineral, Conselho de Minas e Energia, Conselho Nacional de Águas e Energie Elétrica, Conselho Nacional de Petróleo and the Conselho de Exportação de Materiais Estratégicos.

The Sociedade Brasileira de Geologia was founded in 1946 and now publishes, in association with the Conselho Nacional de Pesquisas, the Revista Brasileira de Geociências.

From the seventeenth century to 1970 there were 6,581 contributions to the literature of the geology of Brazil, of which 4,937 were issued in the period 1900-1970. A major bibliography was issued in 1945 (Iglesias and Meneghezzi) and regular supplements have been issued ever since. A summary of all the articles up to 1975 in the bibliography is in the final stage of compilation under the title "Projeto Bibliografia Brasileira de Geologia". A review of geotechnology has been prepared by Guidicini (1976). A number of specialized dictionaries have also been issued (Associação Brasileira de Geologia de Engenharia, 1975, 1976; Guerra, 1966; Leins and Mendes, 1963).

Acknowledgements

We wish to express our grateful thanks to Dr. Fernando Flavio Marques de
Almeida, Dr. Yociteru Hassui and Ms Maria do Carmo S Rodrigues dos Santos
who assisted us during the preparation of this paper.

REFERENCES

ARAUJO, Z. G. de. 1969. Guia de bibliografia especializada. Rio de
Janeiro: Associação Brasileira de Bibliotecários, 208pp. Coleção
Didatica 3.

ASSOCIAÇÃO BRASILEIRA DE GEOLOGIA EN ENGENHARIA. 1975. Glossário de
termos técnicos de geologia de engenharia: Geofísica. São Paulo, 10pp.

—————— . 1976. Glossário de termos técnicos de geologia de
engenharia: Geologia geral. São Paulo, 26pp.

—————— . 1976. Glossário de termos técnicos de geologia de
engenharia: Petrologia. São Paulo, 20pp.

FONSECA, E. N. da. 1973. Origem, evolução e estado atual dos serviços
de docmentação no Brasil. R. Serv. Publ. (Brasilia) 108: 38-52.

GUERRA, A. T. 1966. Dicionário, geológico-geomorfológico. 2nd ed.
Rio de Janeiro, 411pp.

GUIDICINI, G. 1976. Levantamento bibliográfico de geotecnia e engenharia
geotécnica no Brasil e sua indexação por palavraschaves 1920-1975.
São Paulo: Associação Brasileira de Geologia de Engenharia.

IGLESIAS, D. & MENEGHEZZI, M. L. 1943. Bibliografia e índice da geologia
do Brasil 1640-1940. Boletim Divisão de Geologia e Miner logia 111 :
323pp.

INSTITUTO BRASILEIRO DE BIBLIOGRAFIA E DOCUMENTAÇÃO. 1969. Bibliotecas
especializades brasileiras. 2nd ed. Rio de Janeiro, 605pp. Fontes de
Informação 2.

JARDIM, N. S. 1973. Estado atual da documentação geológica. São Paulo,
16pp.

LEINS, V. & MENDES, J. C. 1963. Vocabulário geológico. 3rd ed. São
Paulo: Ed. Nacional, 198pp.

SANTOS, M. S. & JARDIM, N. S. 1975. Divulgação da documentação geológica.
São Paulo, 15pp. For the 1st Reunião Brasileira de Ciência da
Informação, Rio de Janeiro.

INTERNATIONAL ASPECTS OF GEOLOGICAL DOCUMENTATION

SIMON VAN DER HEIDE

Past Secretary General of IUGS, P.O. Box 379, Haarlem, Holland

Summary: Documentation is an essential activity of international scientific unions. In IUGS the first documentation activity was the International Stratigraphic Lexicon. Efforts of the IUGS Committee on Geological Documentation have been successful in establishing a network of information exchange between the main documentation centres in the world, in coordinating a Multilingual Thesaurus for Geology, and, to a minor degree, in the production of critical compilations. The IUGS Committee on Storage, Automatic Processing and Retrieval of Geological Data has developed essential activity in establishing standards to promote consistency in records of geological data and in testing practicability of new techniques.

Documentation, rightly, is regarded as the backbone of science. As real science and its application has the inherent need to be worldwide, documentation is essentially one of the fields of activity of international scientific unions.

In the case of the International Union of Geological Sciences (IUGS) documentation activities were largely inherited from the International Geological Congress (IGC). In fact, one of the items of the first Congress in 1878 was the production of international geological maps. Though maps are considered not to belong to documentation in the strict sense – and in a number of cases they are a mixture of data and interpretation or even fantasy – it should not be forgotten that they represent indeed a most important aspect of geological documentation.

The first activity in documentation within IGC was the International Stratigraphic Lexicon (Lexique Stratigraphique International), an enterprise undertaken with considerable support from the Centre National des Recherches Scientifiques, after a decision at the 19th Session of the IGC in Algers, 1952. After the foundation of IUGS (1961) the commission was incorporated in the IUGS organization as Subcommission for the Stratigraphic Lexicon and continued its most useful documentation activity which to date amounts to a production of more than 120 volumes. In a number of cases second editions have been issued.

Another, less successful, IGC activity commenced after the 20th IGC Session in Mexico in 1956. It was the attempt to establish an International Geological Abstracting Service (IGAS). Though some commercial publishers showed interest in the enterprise, the enormous costs of a service which aims at having an almost complete coverage, were found to be

too great. Moreover, the number of geologists and geological institutions as compared, for example, with the world of medicine (the IGAS Committee worked in close cooperation with the managing director of
Excerpta Medica) was too low to provide a reasonable market for such an
enterprise. The IGAS Committee, after several years of hard work, had to
give up, but its efforts have been most useful as a basis for some of the
major documentation centres in geosciences. Moreover, its metamorphosed
activites were continued by the IUGS Committee on Geological Documentation which is a direct offspring of the old IGAS Committee.

The group of geologists involved in the IGAS and, later-on, in the
Documentation Committee, had gradually developed the idea that publication of collected abstracts was not the most efficient way in dealing with
the problem of information on geological literature. It saw as its primary goal to promote the development of existing major documentation centres in such a way that an interchange between centres in different parts
of the world would be realized. This would provide for reasonably complete coverage of the literature. The ideal situation would, subsequently, be that titles on different subjects, abstracts of publications and
even copies of the papers itself would be available from such centres upon payment of the production costs. Much of this idea of a network of international cooperation has indeed been realized.

A second goal in the initial stage of the Documentation Committee's
activity was the production of critical compilations which should give
reviews of the last 6 to 10 years complete literature on special subjects. The Committee was convinced that such reviews would cover a most
essential need in the fast   growing amount of publications. Of course,
the Committee never thought of an attempt to cover the whole field of
geology by such reviews. It was perfectly aware of the fact that in many
specialized branches of geology scientists have their own compilations.
The idea was to produce these reviews where needed and, if possible, in
connection with a symposium or a conference. The Committee was fortunate
enough to obtain the full cooperation for this effort from one of the
existing documentation centres who provided the authors, without charge,
with the titles on their particular item. Indeed, a number of reviews
appeared, one of the most successful being the Reviews prepared for the
First Symposium on Gondwana Stratigraphy, published by IUGS in 1970. The
production of review papers however, never became a flourishing enterprise. The Committee came to the conclusion that it is extremely difficult to find active research workers who are willing to devote their
energy and time on writing a nearly complete review of the last 10
years' papers in a particular field. They prefer to write a synopsis on

the topical problems which, of course, is extremely good and welcome, but falls outside the field of documentation.

A third activity was the promotion of national documentation of geological publications in the various IUGS member countries. It was successful in a number of cases. In the Netherlands it helped to re-establish, in 1967, this activity which had been given up shortly after the war. It is clear, however, that such documentation and the publication of it requires a considerable effort in the larger countries. It is probably a field in which an international organization of geological documentation could play an active part.

To the activities already mentioned was added, in the second part of the Documentation Committee's existence, a very urgent aspect: automation. A special Automation Board was established within the group. At about the same time, the ICSU Abstracting Board which had been continuously in contact with the IUGS Committee on Geological Documentation, explored the feasibility of multilingual thesauri in science. As a result of these deliberations IUGS was entrusted with the pilot project of a Multilingual Thesaurus for Geology. In the last 6 years this project, in which more than 10 countries are actively engaged, has become the main activity of the Documentation Committee.

It is the author's opinion that the Multilingual Thesaurus is a most important enterprise. Such an enterprise, however, inevitably hampers the development of other activities which may be expected from a Union's Documentation Committee and the question arises what is the specific task of the Union's Committee on Documentation?

Before answering this question it is appropriate to mention one other field in which IUGS has developed considerable activity which is closely connected with documentation and information. It is the field covered by the IUGS Committee on Storage, Automatic Processing and Retrieval of Geological Data (COGEODATA). Details on COGEODATA need not be given here, because they do not belong to documentation in the strict sense and they have, moreover, been published on several occasions. Important, however, is the way of approach chosen by this group. It has never entered the field of data collection itself, it has tried to establish standards in order to promote consistency in records of geological data, it has tried to test the practicability of new techniques and to assist people in choosing the appropriate ways to deal with their data problems in a realistic manner.

This answers the above question. It means that in the author's opinion the specific task of the Union is more the promotion of new techniques and the assessment of standards rather than data collection and documentation management. Indeed, the strength of the Union is the carrying out of pilot projects, even if considerable time is involved, together with the organization of meaningful meetings.

REFERENCES

BURK, C.F., Jr. Compiler. 1977. Computer-based storage and retrieval of geoscience information: bibliography 1973-75. COGEODATA Newsletter, v. 3, no. 2 : 3-14.

DAVIDSON, D.F. Chairman. 1977. Proceedings of seminar on data storage and retrieval of geological data for developing countries. COGEODATA Newsletter, v. 3, no. 3 : 1-23 (6 papers).

HUTCHISON, W.W. Chairman. 1975. Computer-based systems for geological field data: Can. Centre Geoscience Data. Geological Survey Canada Paper 74-63 : 100 pp. (19 papers).

HUTCHISON, W.W. Chairman. 1976. Proceedings of the symposium on capture, management, and display of geological data, with special emphasis on energy and mineral resources. Computers and Geosciences, v. 2, no. 3 : 275-375 (18 papers).

IUGS. 1970. Reviews prepared for the first Symposium on Gondwana Stratigraphy, Mar del Plata, Argentina, September 1967 : 304 pp.

PUBLICATION PRACTICES

USED BY PROFESSIONAL SOCIETIES

WENDELL COCHRAN

Geotimes

5205 Leesburg Pike, Falls Church, Virginia, 22041, U.S.A.

Summary: Professional societies and their journals, like living organisms, have life cycles; they are born, grow, reproduce, and die. These journals should emulate the practices of commercial journals in order to cope with increasingly rapid change in technology.

Professional science societies and their journals go through a life cycle together, with birth being followed by growth, maturity, reproduction, and inevitable death. For much of this life cycle, society and journal may be almost indistinguishable; to many society members, the journal is almost the only visible evidence that the society exists.

Anyone interested in how societies and their publications develop should consult Science: growth and change, by the 10th Director of the United States Geological Survey, H. William Menard. In that book, Menard traces the emergence and growth of a new science specialty as an interdisciplinary development. As an example (not to be blamed on Menard) assume that a specialty called G1 and another called G2 are so distinct that a scientist specializing in one seldom has any particular knowledge of the other. However, sooner or later a few G1 scientists begin applying their specialty to problems in G2, and a few in G2 begin borrowing from G1 for their research.

At first there is little contact. Then, as their number grows, contact increases and the most informal of Information Exchange groups develop -- G1s write to G2s about their preliminary results, they exchange manuscript drafts, visit each other's laboratories . . .

Next, someone organizes a seminar or workshop in what may now be called G3, with G1 and G2 scientists alike attending. (At first these meetings may or may not be held in conjunction with established technical meetings.) Specialty G3 is beginning to develop.

At this point one of the G3 practitioners may go to a commercial publisher such as Elsevier, North Holland, or Hemisphere, and propose a new journal in the new field. Perhaps more likely, the seminars or workshops lead to the organization of a new society or to publication of a newsletter. As with chicken and egg, it doesn't matter whether society or newsletter comes first as the other is sure to follow.

As specialty G3 grows, its society adds members, the greater number of members means more papers, and soon the newsletter is converted into a technical journal. Of course the society may continue to publish its newsletter as a separate medium for distributing news of social events or business meetings and the like. Also, some societies avoid publication for a very long time; the Geological Society of Washington has held well over a thousand meetings, stressing technical papers throughout a century, but apparently it has never aspired to either a newsletter or a technical journal.

When a society produces a technical journal it will appoint or elect an editor. This editor is of course a thoroughly qualified specialist in the science, probably popular with his fellow scientists, and with enough time, ambition, or public spirit he will add a second profession (editing) to the first (science). He (or of course she) probably has a secretary or spouse from which he can embezzle time for typing, handling correspondence with authors and printers, proof-reading, book-keeping, etc. Further, there will be enough others like him to help judge papers for their scientific and literary merit. All those people -- editor, secretary, and referees or associate editors -- are unpaid for their editorial work; none is likely to be a professional editor in any case.

By and large the system works surprisingly well. The society pays no salaries, and membership dues finance typesetting, printing, postage, and maintenance of the subscription list (or the membership list, which is usually the same or nearly so).

The system works well -- to a point. That point is often reached when the society becomes so large, the papers so numerous, and the journal's pages so many that volunteers can no longer spare the time from their work in science. The adolescent society, reaching maturity, then sets up a headquarters office and starts paying rent, salaries to office managers, book-keepers, and (often as a last resort) editors and print-production experts.

The journal, too, has now reached maturity. It and its
society may reproduce, splitting off more-specialized publications and more-specialized societies. Society and journal may
maintain a very long mature stage, or they may slowly die, or
they may be killed.

The shift from free offices and volunteer labour is often
traumatic. Members' dues go up and the members may suspect
that they are paying for a journal that was once 'free'. The
society may soften the blow by selling advertising space and
by raising subscription prices to non-members such as libraries. Those measures may suffice.

However, in the United States, some long-established societies fear death. They fear it at the hands of the US Postal
Service and the Internal Revenue Service (IRS). The IRS con-
tends that differential pricing (charging a non-member more
than a member for a subscription) means that the tax-paying
public is subsidizing the non-profit society and its members.
(A non-profit society not only may not pay taxes; it may also
be eligible for a special postage rate. Loss of its tax-exempt
status means both taxes and higher postage rates.) Following
that line of reasoning, the IRS has moved to revoke the tax-
exempt status of the American Chemical Society, which has perhaps 150,000 members. And in this case the IRS wants to collect taxes back to 1971. Of course the ACS is contesting the
decision, and if it loses in court it may still prevail by way
of a special Act of Congress.

Meanwhile, smaller societies must prepare for the worst.
For example, Geotimes has converted from 'free' distribution
to members of the American Geological Institute's 18 member
societies: all its subscriptions are now paid for by individuals or organizations. However, it is still open to the
charge of differential pricing, as members of the 18 societies
may subscribe for $8 a year although non-members must pay $12.
As a result the staff is seriously considering the impact of
charging all types of subscribers the same rate.

Another route is technological innovation. The Geological Society of America is converting its Bulletin to short summary articles, with full texts available only on microfiche.
The American Association of Petroleum Geologists uses Optical

Character Recognition (OCR), so its authors must use type-writers with a machine-readable typeface. Other organizations plan to use video-display terminals for electronic editing in order to reduce editorial time.

Despite technological progress, a major problem remains for many journals, especially those that have not attained maturity. They may indeed be science journals, but may not be scientific. That should be no surprise, as in the normal course of events a journal's staff may take years to bring its editorial development up to the quality of its science content. The reason is that most of us find it quite hard enough to master one profession, let alone two.

Nor is _Geotimes_ an exception, for when things begin to go awry with that magazine the reason is usually that someone has been re-inventing the wheel -- being a geologist and not an editor and failing to use solutions and methods developed long ago by commercial journals.

In fact the science journals should do even better than the commercial publications. The professional science societies have a great advantage in that they have a great reservoir of built-in loyalty from their members. For example: how many subscribers to _Time_ or _Newsweek_ are loyal in the same way that members of the Paleontological Society are loyal to the _Journal_ of _paleontology_ ?

Even so, more changes are coming. Printing technology is changing so rapidly that (as only one example) ink-jet is becoming a major printing process almost unnoticed by non-specialists. Many publishing specialists fully expect that by 1985 the conventional typewriter will be all but obsolete. Prices of electronic editing equipment seem sure to fall as sharply as have the prices of pocket calculators.

That is not to say that cheap technology will be enough to keep science publications alive, well, and evolving. As Lord Rutherford once said, 'We haven't the money, so we've got to think.'

MENARD, H. WILLIAM. 1971. _Science_: growth and change. Cambridge, Mass.: Harvard University Press, 215 pp.

INDEXING AND ABSTRACTING : DEGREES OF FREEDOM

J. GRAVESTEIJN

Documentation Department

Bureau de Recherches Géologiques et Minières, Orléans (France)

Summary : The introduction of computer techniques in the sixties and the development of the on-line mode have had a great impact on bibliographic information organization in all fields of science and technology.

The existing information systems designed for the production of printed bibliographies and catalogue cards had to be adapted to the new technology or new systems were developed.

In general, the systems became more powerful and the retrieval more pertinent but in all cases the informations scientists operated with systems which involved various kinds of constraints.

In this respect the economic, social and technical aspects are to be taken into account.

Integration of indexing and abstracting techniques will have a decisive impact on future development in the scientific information field.

It is indispensible to promote further cooperation between geoscientific information centers as to achieve the final objective of improved service for the user.

1. The introduction of computer technology has completely transformed the world of indexing and abstracting, particularly during the last decade.

The geological Abstracting and Indexing services first engaged the computer around 1965 and the major information centres successively changed or adapted their systems or production schemes applying the computer processing wherever possible. Initially, the computer equipment available made it possible to improve the processing of printed bibliographies including printouts of indexes, classified citations with annotations or abstracts.

Automatic information retrieval became practical after 1970 when the third generation computers with high-speed processing and large storage capacities became available.

Logically, the first technological developments of the sixties mainly had an impact on indexing techniques while recent technological

developments have affected abstracting services. This is a trend which
will continue in the future.

Reviewing the interface between the technological development
and the parameters defining the philosophy and quality of indexing and
abstracting three groups of factors may be distinguished : economic,
human and technical.

2 - The economic parameters are of course critical.

The market for geological information products is very small.
Only in a few cases, do even the primary journals have more than 2000
subscribers. Subscriptions to secondary journals are much less, sales
of 1000 for an indexing journal covering the whole of the earth sciences
being exceptional.

The introduction of computer processing enabled the secondary
journal publishers to lower production costs or to improve service for
the same cost.

Inflation, the economic crisis and competition of more sophisticated products (SDI - Selective Dissemination of Information - and
computer searches) had an unfavorable impact on the sale of the printed
products. In the case of the Bulletin Signalétique - Bibliographie des
Sciences de la Terre approximately 15 % of the customers being lost
during the last four years.

The loss of this income was compensated for by the sale of SDI
profiles and retrospective searches and in many cases we noticed a
direct relation between the increase in the distributed profiles and
decrease in subscriptions to the printed bibliography. The phenomenon
indicates a dangerous trend.

The secondary journal publishers may decide to discontinue
publishing and concentrate on computer searching only. For example the
last cumulative index of Chemical Abstracts has more than 80000 pages.One
can wonder if publication of the cumulative indexes will continue.

Many users, especially in developing countries, do not have
access to the more sophisticated sources of information and could even
lose entirely their access to secondary information sources.

Cost factors : the cost of indexing and abstracting is continually
increasing. Abstract journals,to reduce costs,are obliged to use the
authors' abstracts or to work with scientific abstractors under contract
(or volunteers).

In-house indexing is still common practice and indispensable. All
the operational geological indexing services have been centralized.
Growing national and international cooperation is a direct result of

economic restraint. Several centres with in-house indexing, combine
their efforts to obtain a more cost - effective data base and to
reduce duplication.

The European network for Geoscience Information is a good example
of this approach. The German, Czech, Finnish, Hungarian, Polish,
Rumanian and Spanish geological surveys or geoscience information
services contribute input to the French PASCAL-GEODE system and receive
in exchange printed or machine-readable geoscience information. The
network is based upon decentralized indexing, use of a common thesaurus,
centralized verification and input and decentralized output services.
Only further integration of existing information centres will solve the
economic problem of increasing costs.

3 - The human factors should not be neglected:for both the users' habits
and the producers' will to adapt existing systems to new technologies
are practical constraints to be taken into account.

For example, some individual users of the BRGM card-index system
were disappointed when BRGM introduced the computerized bibliography in
1968. They normally used small sections of the regional or subject
index-cards and the change was not always suited to their problems. The
choice for the information producer is difficult for both social and
economic factors must be considered. In BRGM the indexing and production
system changed completely from a manual card index to a thesaurus-based
automated bibliography which included subject and geographical indexes.

Experience showed that the new system especially when profiled SDI
was available, offers a better service both to the librarian-documenta-
list community and to the end user. The rigid structure of the thesaurus
is not well adapted for the on-line service. Several auxi liary tables
for rock names, palaeontology systematics or mineral names are available
and a question/answer system guiding the user through the thesaurus and
the classification scheme taking into account the thesaurus evolution
is being developed.

In the case of Bibliography and Index of Geology exclusive of North
America the human factor played a different role. The users were familiar
with a well established journal and this tradition continued when new
computer techniques were introduced in the sixties. The application of
on-line retrieval had a greater impact on indexing techniques for the
American Geological Institute (AGI) than the introduction of automatic
production of bibliographics by second generation computers.

The use of natural language terms was a handicap for on-line searching and AGI is transforming natural language indexing into a thesaurus based system. It should be noticed that a large part of the GEOREF thesaurus is devoted to synonyms which have appeared during the last ten years.

In general, a flexible controlled vocabulary to guide the indexer and user improves the search performances, which unfortunately are often based to a large extent upon frequency criteria.

4 - <u>Technical aspects</u> - To understand the interface between technological development and the information factors it is necessary to realise the structure of the geosciences.

Geology is often considered a natural science or is defined as an observational science or even as a hard science. Actually, the geosciences are situated between biology and physics and chemistry.

4.1- For the geoscience information specialist or system designer, the important factor is the <u>predominance of hierarchically ordered systematic fields</u>. In more than 75 % of the retrospective searches, BRGM has conducted during the last 10 years, geography is a search element. Stratigraphy, rock names, mineral names are also essential search criteria.

This figure is different for SDI service where more current research is involved and methodological aspects are often requested. But even in the case of SDI, 25 % of all profiles contain a geographic element.

In the GEODE-PASCAL thesaurus the number of descriptors covering systematic concepts is 1300, out of a total of 3000, (45 %, geography included). For GEOREF, the percentage is even higher but the subject thesaurus is not yet fully completed. 30 % of Geosystems' main classification headings cover systematic subfields (and in this case, geography and stratigraphy are not included in the classification scheme but are treated in separate indexes).

It is obvious that this predominance of systematic fields has an impact on the indexing language, the file organization and the presentation of printed indexes.

Some examples may illustrate the importance of the systematics for geoscience information.

77

4.1.1- Publications

In the AGI publication. Bibliography and Index of Geology, the geographical and subject entries are combined in one alphabetically ordered index. It is as if the publishers wanted to stress the importance of the geographical descriptors. The second and third level subject descriptors are repeated under the geographical heading. Bulletin signalétique - Bibliographie des sciences de la Terre publishes separate indexes for subject headings and geography, but in the geographical index the set of subjects descriptors is displayed under the geographical heading.

In Geotitles Weekly the citations are presented in classified order with separate geographic and stratigraphic indexes. Geoabstracts - A, Landforms and the Quaternary, publishes also a special geographical index.

4.1.2- File structure, thesaurus

In the case of GEODE, systematic fields are processed through hierarchical internal codes allowing the ordering and grouping of related terms in the published version and for searching.

The non-systematic descriptors are organized in groups (for example properties, methodology) or processed as individual terms (for example off shore, diagenesis, migration, etc).

5 - Language barriers and indexing

There are language barriers, especially in international networks, operating in a multilingual environment.

The European geoscience information network covers eight linguistic communities. The French thesaurus was therefore translated into the various languages(although the abstracts are in French.)

The linguistic problems with regard to output have been resolved for the printed indexes through the publication of bilingual indexes (French and English). In addition the automatic translation of the descriptors in machine - readable form enables the various partners operating the computerized information system to search in their own language.

In the European network the various institutions have a single method of indexing.(Gravesteijn, 1974)

In considering international cooperation beyond the range of a given information system attempts to link different indexing systems operating in different languages pose many problems.

An attempt is being made under the auspices of the International Union of Geological Sciences and the International Council of Scientific Unions - Abstracting Board to design a multilingual thesaurus to link the various systems operating in the five major world languages (English, French, German, Russian and Spanish).

The basic philosophy is that each linguistic group develops its own thesaurus but that connections are established between the selected indexing terms. The core thesaurus has been completed and the first tests will be conducted in 1979. World wide international cooperation will be conditioned by the application of this multilingual thesaurus.

However, a homogeneous data base can only be developed through centralized verification of the indexing. Decentralized handling of information by the various regional information centers without centralized verification produces differences in the quality of the file. The most important constraint for national and international cooperation is that the data base must be homogeneous. Close contacts between the individual indexers and permanent control of indexing are necessary conditions for adequate and efficient searching. The European geoscience experience shows that this need for a common indexing process and centralised input is compatible with a flexible decentralized organization. The common thesaurus can only be developed through joint discussions and decisions.

6 - Indexing - abstracting

When considering the future development the importance of intregrated indexing and abstracting must be stressed. The storage and access capacities of the latest generation of computers make it possible to store citations, indexing terms and abstracts. A search with controlled indexing terms can be refined through the use of abstracts in natural language. This means that the conventional use of abstracts as a selection tool for reading has now a computerized parallel.

In a BRGM inquiry on whether abstracting of a given article is necessary, it was shown that for more than 50 % of the literature, indexing terms only partially cover the information contained in the documents. This was especially true for papers dealing with regional geology and containing many aspects of the earth sciences. These papers need more analysis than just in - depth indexing. For articles containing numerical data in the fields of mineralogy or mineral economics, abstracts can give additional information. For other fields the indexing terms are sifficiant for efficient retrieval (Gravesteijn, 1975).

In France and other European countries automatic translation of scientific abstracts is the subject of much research. Several methods are being explored and the most promising is the use of slightly formalized abstracts, excluding certain grammatical structures such a passive and impersonal forms.

These approaches could also lead to the automatic analysis of phrases contained in the abstract and extraction of controlled significan indexing terms.

Two possible applications of integrated abstracting and indexing are the creation of multilingual data bases and the publication of bibliographies with multilingual indexes and abstracts. This is especially encouraging in the European multilingual environment.

REFERENCES

GRAVESTEIJN, J. 1974. Geological Information - An example of international cooperation. In : Proceedings of the ICSU AB General assembly meeting in Berlin. Paris : ICSU AB : 221-223.

GRAVESTEIJN, J. 1975. Abstracts in the GEODE system. Communication presented at the discussion meeting of the Geological Information Group of the Geological Society of London on November 6, 1975. (not published).

THE ROLE OF THE UNISIST/ICSU-AB REFERENCE MANUAL

IN DATA BASE PROCESSING

John Mulvihill

American Geological Institute

5205 Leesburg Pike, Falls Church, VA 22041

Summary: Most computer-produced bibliographic data bases must serve two masters, (1) printed products such as abstract bulletins, bibliographies, and indexes, and (2) on-line and batch searching. It is also desirable that a data base be compatible with others in format and data elements. The Reference Manual is judged on these criteria, based on its use for the GeoRef data base.

The Reference Manual for Machine-Readable Bibliographic Descriptions (Martin, 1974) is an international standard for the data elements and format of bibliographic data bases in computer-readable form. The Reference Manual is used for the GeoRef data base.

The tape format called for in the Reference Manual is a specific implementation of the international standard, ISO 2709 Communication Format for Bibliographic Records. This same ISO Standard is also the basis for the Marc format in use at the Library of Congress.

The Reference Manual was prepared by the UNISIST/International Council of Scientific Unions-Abstracting Board (ICSU-AB) Working Group on Bibliographic Descriptions, with assistance from the ICSU-AB member services. The membership of ICSU-AB includes many of the large abstracting and indexing services of the world.

The UNISIST International Centre for Bibliographic Descriptions (UNIBID), located in the British Library, has been established to maintain and revise the Reference Manual.

Representatives of ICSU-AB, the National Federation of Abstracting and Indexing Services (NFAIS), the Association of Scientific Information Dissemination Centers (ASIDIC), and the European Scientific Information Dissemination Centers (EUSIDIC) have used the Reference Manual as the basis for several joint discussions of a standard for bibliographic data bases, from which discussions have come recommendations for revision of the Reference Manual.

In spite of such support the Reference Manual has not been widely adopted. There remain nearly as many tape formats in use as there are data

bases. GeoRef is one of the few data bases to have fully adopted the
Reference Manual. The consequences of using the Reference Manual are
reviewed below, as illustrated by GeoRef, in relation to printed
products, searching, and compatibility with other data bases.

PRINTED PRODUCTS

If a data base were to be used only to produce one bibliography with
author and subject indexes, then only five data elements might be considered adequate: author, title, source, index terms and accession
number. However, if there is a requirement to produce an annual index for
a journal from these data elements, it cannot easily be done. To produce
an annual index there is the need to be able to refer to the journal
articles by page number. But page number is embedded in the source field
and consequently cannot be handled by computer. There might also be a need
to address the journal name, volume and issue numbers, and publication date,
as units. These too are embedded in the source field. Each different type
of publication from the data base makes new demands.

The Reference Manual provides for these demands by specifying that
bibliographic data be subdivided into a number of discrete, well-defined
data elements. This makes it possible for the computer to sort, alter and
assemble this data to meet varied print formats.

The problem of preparing varied printed products from a data base is
compounded if the data base includes several document types such as book
and serial. Then the data elements take on a dual function; for example,
the title field may be either a book title or a journal article title.

A further complication arises if the data base includes references on
multiple bibliographic levels, e.g. book chapters, books, and collections
of books. Then the title field may be: a chapter title, a book title, a
collection title, or a combination of these. The same complexity applies
to the author and source fields.

The Reference Manual handles the complications introduced by the
presence of multiple document types and bibliographic levels by prescribing
that type and level be specified for every reference entered into a data
base. Each reference must be designated as one of the following document
types: book, serial, report, thesis, or patent. Other types can be
accommodated. Similarly, each reference must be assigned one of three
bibliographic levels: analytic, monographic, or collective. For example,
a journal article would be designated serial-analytic. The matrix in
Table 1 illustrates the possibilities.

Table I, DOCUMENT TYPE BIBLIOGRAPHIC LEVEL MATRIX

	Analytic	Monographic	Collective
Serial	✓	✓	
Book	✓	✓	✓
Report	✓	✓	
Thesis		✓	
Patent	✓	✓	

(Table 1 is from the Reference Manual (Martin, 1974:10)

The Reference Manual makes type and level the key to the selection of data elements for each reference. From the type and level selected, the Reference Manual designates which data elements are essential for the reference (see Table 2).

The notions of document type and bibliographic level have been common in libraries for centuries. However, correlation of the two and selection of data elements based on this correlation are not common in data bases. In many data bases there is little distinction between the various types and levels. The Reference Manual makes such distinctions clear. For example, it has three person fields: person-analytic, person-monographic, and person-collective, based on type and level.

The concepts of type and level facilitate the building of detailed, understandable printed references. The recently revised American National Standard for Bibliographic References (1977), has adopted these concepts as found in the Reference Manual, making the two standards compatible in this fundamental approach toward bibliographic description. Consequently it should be possible for a data base using the Reference Manual for bibliographic references in computer-readable form to use the American National Standard for bibliographic references in human-readable form.

The specificity of the Reference Manual requires that a data base have many data elements. For example, GeoRef has a total of 70 data elements, 35 from the Reference Manual and 35 for pre-Reference Manual data and for types of data not covered in the Reference Manual, such as index terms, medium, map scale, and geographic coordinates. Nearly all bibliographic data for current references is in the 35 Reference Manual data elements (see Table 2).

A variety of special bibliographies, year-end indexes and current
titles lists have been photocomposed from GeoRef since the American
Geological Institute adopted the Reference Manual data elements. In no
case have we been restricted in formating a bibliography or index due to
limitations imposed by the data elements.

The multiplicity of data elements in the Reference Manual has
discouraged its use. But this need not be so, since for each type and
level the Reference Manual specifies the set of data elements which are
essential. Choice of type and level substantially reduces the number of
possible data elements for any single reference. For GeoRef there is an
index form and corresponding cathode ray tube screen for each type-level
combination, in which only the data elements required by that combination
appear. The indexer or typist working on any given reference need be
concerned only with the data elements appropriate to that reference.

SEARCHES

The same characteristics which make a Reference Manual data base
suitable for printed products also make the data base a good one for
searching.

Specific, well-defined data elements and the use of type and level
enable a search service to provide multiple access points for searching a
Reference Manual data base. For example, in GeoRef the following data
elements are separately and directly searchable on-line through System
Development Corporation:

> Document Type and Bibliographic Level, Accession Number,
> Entry Year, Title, Author, Affiliation, Corporate Author,
> Source, Coden, ISSN, ISBN, Publication Year, Century,
> Report Number, Language, File Segment, Availability,
> Category Code, Index Term, Index Word, Supplemental Term,
> Geographical Coordinates, and Annotation.

Document type and bibliographic level enable the searcher to limit
a search to any combination of the two, such as book-monographic. By
specifying book-monographic the title becomes a book title only and the
author becomes book author, editor, compiler, etc. only.

In addition, the format of the bibliographic record facilitates
searching and conversion to other formats. The ISO bibliographic record
is divided into three sections: a fixed length leader, a variable length
directory, and data fields of fixed or variable length. This format is
suited to higher-level language processing, e.g., it can be processed
using COBOL, PL/1, and BASIC. Consequently it is easier to program for

and more machine-independent than if it required assembly language programs.

The directory makes it easy to pick out data from a record for conversion or for searching.

A deficiency in the ISO standard and consequently in the Reference Manual is its failure to specify a method for spanning physical records when the logical record exceeds the established record length of the data base, but on the whole the format is well-suited to be a distribution format for bibliographic data bases.

COMPATIBILITY

Consider the problem of a search centre handling multiple data bases. Somehow it must deal with the idiosyncrasies of each data base. Typically the centre devises its own format and writes a conversion program to change the format of each data base received to its format. Each data base requires a different conversion program. On receipt, each data base update is converted to the centre's common format before it can be used. Each time a data base alters its data elements or format, its conversion program must be altered and tested. Further, although the search centre adopts a common format, the data elements of the various data bases it searches are not uniform and can be only imperfectly standardized. One data base will mix corporate authors and personal authors, another will separate them, etc. Consequently, the centre cannot fully reconcile the differences between data bases.

In a similar fashion, when data bases seek to exchange or pool references, conversion programs and processing are required for each data base involved.

The searcher must be aware of the idiosyncrasies in each data base which remain in the search centre's common format. The searcher loses time, money and search relevance and recall because of the current diversity among data bases.

If a single well-designed standard were adopted by all data bases, then the above problems would be greatly reduced.

A distinction needs to be made here. Existing data bases would not need to discard their formats to achieve this standardization. They could have their own internal formats, plus a standard distribution format such as the Reference Manual, generated from their internal format.

GeoRef, for example, is updated and maintained at the American Geological Institute in a non-standard internal format from which the printed products are photocomposed. However this format is converted

to the Reference Manual format for distribution and searching. The
internal format is more suitable for data base maintenance and photocomposition than the Reference Manual format. Conversely the Reference
Manual format is better for searching and easier to convert. But the
data elements in both formats used for GeoRef conform to those of the
Reference Manual.

SUMMARY

Advantages of Reference Manual:

1. Endorsed by UNISIST and ICSU-AB
2. Maintained by UNIBID office at British Library
3. Compatible with ANSI standard for printed citations
4. Tape format based on ISO 2709 Standard
5. Incorporates other International Standards:
(Transliteration, Country Code,
6. Requires document type and bibliographic level
7. Specifies format and content of each data element

Current problems with Reference Manual:

1. Not widely used
2. A revision of the Reference Manual is needed

CONCLUSION

As to the future, one would hope that the Reference Manual format
will be adopted as a distribution format by new data bases and that
established data bases will switch to it when they reach a point where
a change in their data elements and formats is needed.

If this happens, data base searching would be much improved, and
millions of dollars would be saved annually in programming, conversion,
and search costs. Perhaps it is this economic pressure which will provide
the needed impetus toward adoption of a standard distribution format such
as the Reference Manual.

References:

American National Standard for Bibliographic References. 1977
New York: American National Standards Institute, 92 p.
(ANSI Z39-29-1977)

MARTIN, M. D. Compiler. Reference Manual for Machine-Readable
Bibliographic Descriptions. 1974. Paris: UNESCO, 71 p.

Table 2, DATA ELEMENT MATRIX

Tag	Field name	Serial		Book			Report		Thesis	Patent
		A	M	A	M	C	A	M	M	A/M
AØ1	International Standard Serial Number (ISSN)	E	E							
AØ2	CODEN (interim alternative to ISSN)	*	*							
AØ3	'Short title' of serial	E	E							
AØ4*	Series designation									
AØ5	Volume number	E	E	E[1]	E[1]					
AØ6	Issue or part number	E	E	E[1]	E[1]					
AØ7	Other identification of issue or part	E	E							
AØ8	Title of contribution (analytic)	E		E			E			
AØ9	Title of volume, monograph or patent document		E	E	E		E	E	E	E
A1Ø	Title of collection			E´	E´	E				
A11	Person associated with a contribution	E		E			E			
A12	Person associated with a monograph		E	E	E			E	E	
A13	Person associated with a collection						E			

. For books (at analytic and monographic levels) fields AØ5, AØ6 and A1Ø are essential only if the item is part of a collection having numbered parts.
* Tags marked with an asterisk indicate data elements which are never designated as essential.

Tag	Field name	Serial		Book			Report		Thesis	Patent
		A	M	A	M	C	A	M	M	A/M
A14	Affiliation - contribution	E		E			E			
A15	Affiliation - monograph		E							
A16*	Affiliation - collection									
A17	Corporate author - contribution	E		E			E			
A18	Corporate author - monograph		E		E			E		
A19	Corporate author - collection					E				
A2Ø	Page numbers	E	E	E			E			
A21	Date of issue or imprint	E	E	E	E	E	E	E	E	
A22	Date of publication[2]									E
A23	Language(s) of text	E	E	E	E	E	E	E	E	
A24*	Language(s) of summaries									
A25	Publisher: name and location (monograph or collection)			E	E	E				
A26	International Standard Book Number[3] (ISBN)			E	E	E				
A27	Edition			E	E	E				

2. Field A22 may be used for any literature type where the actual date of publication is known to differ from the nominal date of issue.
3. Field A26 (ISBN) may be used for any type of literature if the publisher has chosen to assign an ISBN to the piece being recorded.
* Tags marked with an asterisk indicate data elements which are never designated as essential.

Table 2, DATA ELEMENT MATRIX (cont.)

Tag	Field name	Serial		Book			Report		Thesis	Patent
		A	M	A	M	C	A	M	M	A/M
A28	Collation: description of non-serial collection					E				
A29	Collation: description of monograph				E	E	E		E	E
A30	Name of meeting[4]									
A31	Location of meeting[4]									
A32	Date of meeting[4]									
A33	Identification of patent document									E
A34	Person associated with a patent document									E.
A35	Corporate body associated with a patent document									E
A36*	Domestic filing data									
A37*	Convention priority data									
A38*	Reference to a legally-related domestic document									
A39	Report number						E	E		

4. Fields A30, A31 and A32 are essential - regardless of literature type - if and only if the piece is formally designated as constituting the published proceedings of a meeting.
* Tags marked with an asterisk indicate data elements which are never designated as essential.

Tag	Field name	Serial		Book			Report		Thesis	Patent
		A	M	A	M	C	A	M	M	A/M
A40*	Name of performing organisation									
A41	University (or other educational institution)								E	
A42*	Degree level									
A43	Availability of document						E	E	E	
A44*	Source of abstract									
A45*	Number of references									
A46*	'Summary only' note									
A47*	Abstract number(s)									
A99	Ancillary data									

* Tags marked with an asterisk indicate data elements which are never designated as essential.

Table 2 is from the Reference Manual (Martin, 1974:5-6)

COMPUTER-ASSISTED EDITING OF BIBLIOGRAPHIC INFORMATION

G. N. Rassam

American Geological Institute

5205 Leesburg Pike, Falls Church, VA 22041

Summary: The role of the computer in the decision-making process
editing bibliographic information is described. It is shown that automatic
processing, using several data files, can reduce the time needed for the
editorial function, and provide many valuable checks and controls on the
flow of information between input and output. The functions of computer-
assisted editing are described for the various data elements in the
bibliographic citation. Particular attention is given to the role of
computers in editing indexes.

> "We decided to pick up with fashion," confessed Vladimir
> Vibrobov, chief engineer of the electrical manufacturing
> plant in Minsk. "We thought that if it was a thinking
> machine, it would think for us. And it only brought us
> trouble." The Washington Post (March 15, 1978)

Computer-aided editing is primarily the function of the ability of

modern computers for handling large amounts of data in a short time.

It can be demonstrated, in fact, that an editor aided by the computer

can produce about twice as much work as one relying on more traditional

methods.

Furthermore, a computer can provide often very useful additional

information that can help in the management of a data file. Such informa-

tion of a statistical nature--production figures and index term frequency--

can provide the basis for rational decisions.

In bibliographic work however, there remains subjective or not easily

standardized information (such as author names or titles) that requires

human intervention on a routine basis. In addition, the standardization

necessary for computer manipulation means that such index phrases as:

Jurassic through Tertiary, although quite descriptive and meaningful, can

no longer be used. The substitute, Jurassic, Cretaceous, Tertiary, is

pallid in contrast and may have a totally different meaning.

A computer cannot distinguish between homonyms nor can it handle

negative concepts with efficiency: not Tertiary or pre-Mesozoic are not

easily amenable to mechanical verification.

In the early development of computers, they were simple machines

requiring a great deal of standardization on the part of the user. As the

computer has become more sophisticated and therefore more flexible, the

options available to the user have increased and there is less demand for standardization. Thus, the battle for standardization (efficiency vs. improvisation) has been a dynamic, even dialectic, process.

Present changes in computer technology are proceeding in the direction of less reliance on paper printouts and more on on-line editing, such as a system using a minicomputer with a cathode-ray tube (CRT) terminal.

In such a system a set of screens can be programmed to follow the specific parts of the UNISIST standards described below. (Martin, 1974) The screens then can be used for data input on the CRT terminals and the information stored on disks. The minicomputer processes the input. An editor recalls the processed material on the screen of his terminal, and selects the desired documents to be edited. Essential editing functions such as deleting words or lines, inserting words or lines, substituting words or lines for existing ones, and other changes on the text, can be made directly and with the minimum of effort.

Validation can be made quickly and simultaneously with data entry and the results (error messages and the like) can be displayed on the screen, providing the editor with the opportunity for instantaneous decision-making.

Machine manipulation can scan several files to bring in information or delete or change it; can control the punctuation and sequence of a bibliographic citation; and can provide economic shortcuts for many routine problems. What it cannot do is think or rather it cannot create. Even at its sophisticated best (CRT terminals), it still cannot provide the psychological satisfaction of writing on a solid medium such as paper. Electronic writing is not a perfect substitute but the typewriter was probably not considered as a much better substitute for handwriting.

Bibliographic Information:

In editing bibliographic material, the editor is faced with several decisions after the basic question of inclusion or exclusion of material is decided. These decisions relate to the scope of the documents involved; the classification of the documents; indexing and final editing.

Mechanical manipulation of data helps primarily in the third decision-making process and this paper will delineate some of the possible areas of utilization of such manipulation as well as its limitations. The structure will follow the basic data elements of a bibliographic citation.

1.   Identification numbers: Documents can be identified in various ways and for various reasons. Computers can help in either automatically adding sequential numbers to documents, or verifying that the number is of a certain length, is numeric only or not, and that its length is of a certain pre-determined magnitude.

2.   International Standard Serial Number (ISSN): Verification can
be made of the presence of numerals only, length, and validation by
check digit or look-up in a separate serials file.

3.   CODEN (journal designation): Verification of upper-case letters
only, length, and validation by check digit. Presence of either ISSN
or CODEN is mandatory for serials in some bibliographic systems, for
example, and can be ascertained by the computer. The abbreviated
form of the serial title (short title) and the full title can be
stored in correspondence with the given CODEN or ISSN on a separate
serials file, and inserted into a reference automatically.

     This file, with such additions as frequency of publication and
priority of coverage, can be used to alert the editor to gaps in
coverage as well as to changes in the serial title. If a CODEN or
ISSN is not on the serials file it can be rejected by the computer.

4.   Document designation: Whether a given document is a serial or
a book or a report or any other kind of document, and, in addition,
whether the given document is on the analytic level or the monographic or the collective can determine the sequence and kind of
information included in the bibliographic citation. Such sequence
can be automatically controlled. Thus a serial analytic document
(the most common type of document published, estimated to comprise
about 80% of all types of printed documents in the geological field)
would require, according to the UNISIST Manual, the existence of an
ISSN or CODEN, a "short title", a volume number, an issue or part
number, a title, an author, page numbers, date of issue and language
of text. The presence of each of these data elements in the
reference can be checked and its contents verified.

5.   Identification of volume, issue, or part: The length and kind
(numeric or non-numeric) of this information can be controlled.

6.   Title: The presence or absence of levels of titles (analytic,
monographic or collective), can be ascertained. The contents of
this element are most variable and are therefore least amenable to
automatic manipulation.

7.   Authors and/or other persons associated with a document: A
person's name is the most subjective part of a bibliographic file. The
variety in usage and form is infinite. The problem is compounded in
the case of an international file. The computer-provided assistance
is limited to the compilation of authority files, author lists

(deduplicated to facilitate perusal) and sorted lists.

The editor's task is facilitated by such devices but many of the problems involved can only be solved by editorial judgement. Compound surnames, prefixes like van, von, van der, della, al, el, and authors who change their names all contribute to the variability that must remain as a challenge to the human processor.

8. Affiliation and corporate authorship: Similar problems pertain here to those of individual authors but the use of standard country codes helps in shortening the time spent. Country codes can be easily validated by computer which can add the properly spelled country name to the record at the same time.

9. Other bibliographic data: Other information such as pagination, languages and illustrations can be, to various extents, standardized and therefore automatically controlled.

10. Indexing: In general, the computer at different stages of the processing and editing process can provide verification and consequent error messages alerting the editorial staff to mistakes. Computer-assisted editing is most effective, however, in the area of editing the indexes of bibliographic citations.

Index terms which are used in a formal indexing structures can be tagged and entered in the record. Should the individual term be required for use in more than one sequence, the designation of the index term can be used instead of writing the term several times. An example:

        A.   Algeria

        B.   Economic geology

        C.   Petroleum

        C.   Petroleum

        D.   Occurence

        E.   Reservoir rocks

        F.   Triassic

        G.   Stratigraphy

        H.   Stratigraphic traps

The repetition of the term petroleum in this example need not entail writing the word twice; repeating the letter 'C' in the proper sequence will suffice. Similarly, other components of the bibliographic citation can be entered as letter or number designations saving the indexer considerable time. An example is the use of coordinates (longitude and latitude) for geographic descriptions. Once the different regions are assigned their coordinates in a separate file, then the entry of an area term or a letter substitute automatically triggers the computer to assign the appropriate coordinates to the document.

The processing can define the level and sequence of index terms, length of terms, whether the term is valid or not (in comparison with a thesaurus file), whether a term is to have cross-references or not (by comparison with a cross-reference file), and whether a term is to have an autoposted term or not (by comparison with an autoposting file):

a.   Level and sequence of index terms

Index terms may have a heirarchical structure which can be ascertained automatically. Thus, in the example given previously, where a three-level heirarchy prevails, the index terms can be checked and verified to belong to one or more pertinent levels. Algeria, for example, can be allowed on the first (highest) level and on the third but not on the second.

The sequence can be pre-determined by the tags assigned to the terms but the sequence in the sense of a logical progression of terms cannot be pre-defined; i.e., if the term "stratigraphy" is allowed on the second level and is entered by mistake on that level instead of "Triassic" in the given example, then such an error has to be detected by the editor. In addition, contextual differences in the usage of index terms are difficult to process automatically, especially if the terms were used on the same level or have the same apparent weight. Thus the term igneous rocks used in the meaning of all igneous rocks is often hard to differentiate from the usage of the term as a "grouper" of only few types of igneous rocks.

What is even more difficult is the question of syntax and its manipulation by machines. Not much work has been done with geologic data bases to delineate the possibilities here.

b.   Validity of index terms:

The validity of index terms can be verified by making a comparison with an established thesaurus file. Invalid terms are then flagged and distinguished from valid ones.

This process greatly aids the editor as it isolates invalid terms due to misspelling (which can then be corrected), due to mistakes in capitalization, or due to exclusion from the thesaurus (which gives the editor the option of including them, validating them, or keeping them excluded).

c.   Cross-referencing index terms:

A file containing "use", "use for" and heirarchical relationships can provide cross-references. Thus in the example given, the presence of underlined petroleum on the 3rd level can trigger a cross-reference of the type: petroleum see also under economic geology under Algeria.

d.   Autoposting of index terms:

Systematic index terms fitting in a general classification can have various levels of broader terms automatically posted to increase the efficiency of automatic searching without spending the time of the indexer on such essentially repetitive tasks. Examples would be stratigraphic terms (Maestrichtian getting Cretaceous autoposted); geographic terms (France gets Europe); and mineral terms (galena gets sulfides and minerals). The limitations here are in the inherent difficulties in classifying natural objects such as igneous rocks or minerals. Should a mineral belong to two chemical groups, such as "tsumebite" belonging to both sulfates and phosphates, it can have the group names autoposted, but if the mineral is classified in some individual paper as a sulfate then it would still autopost phosphates with the resultant loss in specificity.

Autoposting in the other direction, i.e., a broad term giving all narrower terms, can be done also but is of no practical value here.

11. Timing: The various checks described above can be applied selectively and at specified intervals to provide efficient means of controlling processing and editing. Verification of the data elements can be made continuously at each step of the processing cycle with each correction and each addition marked. Processing, on the other hand, can be defined by stages each of which is represented by a computer printout emphasizing different aspects of the material. An example within a month's cycle of work would be lists of authors, sources, index terms, etc.

In general, this function of computers is often overlooked or not emphasized enough. The fact that it is possible with the computer to break the production cycle into discrete steps each with its own

functions and checks, means a great advantage over manual systems
of control where the flow of work is more continuous, thus more
arduous, necessitating record-keeping of vast magnitudes for large
files.

In conclusion, it should not be too difficult to use the computer as
a tool for greatly reducing the time devoted to routine editorial chores
and as an editorial management tool for planning, all the time keeping in
mind that scientific and stylistic ambiguities need not be completely
ignored or eliminated.

The choice of machine vs. human or various mixtures thereof can be
made on the basis of efficiency and economy which, in a broad sense, do
include the total environment in which an editor works and should take
into consideration such factors as user demand and preferences.   The
computer provides the editor with an extra tool for that task.

Martin, M. D. 1974.   Reference Manual for Machine-Readable Bibliographic
    Descriptions, 71 p., UNESCO, Paris.

GEOLOGICAL DATA MANAGEMENT

KEITH G. JEFFERY

NERC Central Computing Group

Rutherford Laboratory, Chilton, Didcot, Oxfordshire, OX11 0QX

Summary: One of the major objectives of the science of geology may be
regarded as the understanding of geological processes. A long period of
evolution and learning by many people in different countries of the world
has provided simple computer systems for handling geological data. The use
of sophisticated computer systems to handle geological data with the aim of
understanding these processes is just starting. The different techniques
and approaches are discussed, with particular reference to database concepts
and the G-EXEC system.

Introduction

The discussion is centred on systems which handle raw data, not bibliographic information, or indexes to and abstracts of published documents.
The use of automated data collection equipment in geoscience is increasing
both in magnitude and scope so that data pertinent to the science of
geology are accumulating at an ever-increasing rate. It may be postulated
that in geoscience the real problem is to understand the processes
active in the geological environment and to this end it becomes important
to reduce masses of raw data into fewer variables which can be used to
understand the geological processes of interest.

The use of computer systems as an aid to this understanding may be
considered under three main aspects:

(a) data collection/validation and storage
(b) data analysis (in the widest sense)
(c) simulation

and all three of these aspects require fairly sophisticated data handling
techniques for them to be effective and useful to the geologist.

It appears that geologists frequently handle data without thinking of the
measurements, observations and possibly the thoughts they have as data.
The root of this difficulty seems to be that the processes studied in
geoscience are so complex that the relationship between raw data and the
overall process is obscure. This highlights the need for analysis of the

96

process being studied and the reduction of the process into component
smaller-scale processes until a unit size of process is reached which
can be understood in terms of raw data. This technique is known as
systems analysis. However, a geologist in his everyday work tends to do
this reduction of complexity in his head. Furthermore, he then combines
the understanding so gained of one small part of the system into the model
he holds in his mind of the overall system. The challenge of computer-
based systems to help geologists in their everyday work is that they must
perform this breakdown and re-integration quickly, scientifically,
reproducibly and under control of the geologist.

Database concepts

The basic idea behind databases is that they should provide a store of
data which is under one control (but not necessarily stored all in one
place) and which has only one entry for each item of data; ie there is
no duplication of data. The database has two aspects. The database
content is the values assigned to the data items. The database structure
is the relationships between data items or groups of data items. If the
data content is precise and unbiased then the database reflect a view
of each individual item of data as observed in the real world. If the
structure is also precise and unbiased then the database will also reflect
the relationships between items in the real world as observed and
collected by the database creators.

Two main philosophies of database technology exist at this time; the
CODASYL approach and the Relational approach. The CODASYL approach
(CODASYL, 1971) involves building and processing databases using programs
built by the user and incorporating a sub-schema which is a description
of the content and structure of that part of the database of interest
to the program. The whole database is described by a schema which has to
be defined prior to creation of any part of the database. This, in turn,
means that the database creator(s) must have a fully-planned-out view of
the database before any data may be input. The CODASYL standards were
defined by a committee which had its evolutionary roots in the committee
which defined the COBOL language. The CODASYL recommendations are to some
extent an extension of the data definition and language sections of the
COBOL standards.

The Relational approach has been defined by the work of E.F. Codd
(CODD, 1970) of International Business Machines and the basic idea is
that the database should consist of data items or groups of items, and
inter-relationships between them. The items should mirror the real world
and the inter-relationships should also mirror observed relationships in
the real world. This concept allows for great flexibility in database
structure because predefinition of the database structure is not required.
It also provides for the use of powerful operands (the relational
operators) on the database which provides the user with extremely
powerful data manipulation facilities.

Geological database systems

It is not possible to provide an exhaustive list of systems that have been
developed for handling geological data. However, a general picture of the
development of such systems follows. Some early work on computer handling
of geological data was based on files built by individual scientists,
commonly in university environments. The next stage of organisation is
exemplified by groups of scientists in geological surveys or commercial
organisations who defined the content (and in some cases the simple
structure) of a file and then proceeded to store available data. Such
systems usually had specifically written software for data input and
retrieval and some had additional software for data analysis and graphics.
In Great Britain such systems were produced by collaboration between the
Institute of Geological Sciences and the Atlas Computer Laboratory;
(Gover, Read Rowson, 1971; Gill, 1975) in Canada some early systems
for mineral deposits were of this nature and others have been developed
by the Geological Surveys in West Germany (DASCH), (Mundry, 1973)
and Sweden (GEOMAP), (Berner, Ekstrom, Lilljequist, Stephenson, Wikstrom,
1972).

From such work database systems in geoscience evolved, and such systems
are now becoming accepted. In the USA the GYPSY system handles data with
hierarchic structure while the more recent GRASP system works with data
structured into the linear form of the relational database approach. In
Canada, SAFRAS (Sutterlin, De Planke, 1969) was built for hierarchic data
structures. The commercially developed System 2000 has been used for
hierarchic and network-type structures. In France, SIGMI (Kremer, Lenci,
Lesage, 1973) handles hierarchic data and also provides for more complex

structural links between data items. In Great Britain, G-EXEC has been developed based on relational database concepts.

A useful summary of papers on some of the systems was provided (Hutchison, 1975) at a COGEODATA Conference held in 1973.

G-EXEC

G-EXEC (Jeffery, Gill, 1976) is a machine-independent, generalised data-handling system. On the major British installation well over 1000 mega-bytes of data are stored on demountable disk packs and the system processes some 1300 jobs per month. It handles data from several fields of environmental science in addition to geoscience. In the geological domain it has been used for data on borelogs, geochemistry, wells, mineral assessment, commodity production and trade, petrology, structural geology and geotechnical data. The system has also been used for administrative purposes; management information, telephone directory and accounting systems have all been run. Each user accesses and processes data by simple commands and so the handling of data is made relatively easy.

The aim of such a system is to reduce the mass of raw data available into a manageable amount of facts which can be used by geologists to understand geological processes. G-EXEC currently satisfies much of the user require-ment for data validation, storage, retrieval and analysis (including statistics and graphics). However, these facilities still leave the user with the task of integration of facts and model building. The next step of the use of simulation for geoscience in a database environment is only just receiving attention.

Simulation

There have been some notable attempts at geological simulation in the past, and with increasing knowledge of geological processes and increased availability of computing facilities the simulations have become more complicated and presumable closer to the real world. However, most geological processes are very complex, consisting of complex sub-systems, and the understanding gained from the models may be relatively superficial, although extremely useful.

It is hoped that systematic breakdown of geological processes by analysis
followed by deeper understanding of the component sub-systems through
the use of databases and sophisticated data systems will lead to the
building of simulation models which reflect more closely the geological
processes observed in the real world.

REFERENCES

BERNER, H., EKSTROM, T., LILLJEAUIST, R., STEPHANSSON, O., WIKSTROM, A. 1972.
GEOMAP - A data system for geological mapping. Proceedings of the 24th
International Geological Congress, Section 16; 3-11.

CODASYL. 1971. Programming Languages; New York. Association of Computing
Machinery Data Base Task Group. 269pp.

CODD. E.F. 1970. A relational model of data for large shared data bases.
Communications of the Association of Computing Machinery, 13: 377-387.

GILL, E.M. 1975. Feasibility studies for a petrographical data bank.
Report of the Institute of Geological Sciences, No 75/3. London.
Her Majesty Stationery office. 31pp

GOVER, T.N., READ, W.A., ROWSON, A.G. 1971. A pilot project on the storage
and retrieval by computer of geological information from cored boreholes
in Central Scotland. Report of the Institute of Geological Sciences,
No. 71/13. London. Her Majesty's Stationery Office. 30pp.

HUTCHISON, W.W. 1975. Computer based systems for geological field data
Geological Survey Paper 74-63. Ottawa. Geological Survey of Canada.
100pp.

JEFFERY, K.G., GILL, E.M. 1976. The design philosophy of the G-EXEC System.
Computers and Geosciences. 2: 345-346.

KREMER, M., LENCI, M., LESAGE, M.T. 1973. SIGMI un nouvean systeme
d'interrogation de fichiers oriente vers l'utilisateur. Rapporte
Interne E.M.P. CIG/R73/5.

MUNDRY, E. 1973. Ein dokumentations-und Abfrageprogramm fur
Schichtenverzeichnisse (DASCH). Geol. Jb. A7: 35-50.

ONLINE INFORMATION RETRIEVAL - PRESENT STATUS

AND FUTURE IMPLICATIONS

GRAHAM LEA

Director, Geosystems, P.O. Box 1024, London SW1, UK

Abstract

Geological information workers who are not computer retrieval oriented are victims of Future Shock. Even those who have some background in on-line retrieval may not be prepared for the speed of new developments, particularly in the forthcoming merging of bibliographical and numerical databases. The problems of civilisation, and in particular our post-industrial society, require that we have the wisdom and resolution to act with existing knowledge and information to devise effective solutions.

These problems can be addressed intelligently by improving our information systems, and the communication of our knowledge.

WHAT INFORMATION IS AVAILABLE FOR THE APPLIED GEOLOGIST

MICHAEL HAINES

Charter Consolidated Limited

40 Holborn Viaduct, London EC1P 1AJ

Summary: Often due to exploration budget constraints the applied geologist finds himself a jack-of-all-trades as well as master-of-his-own.
This paper provides a broad-brush picture of the types of information
required and provides some detail in areas where difficulties can arise
in obtaining information.   Whilst recognising there are many facets of
the geological profession, this introduction to the session dwells mainly
on information in the context of the exploration for, and exploitation of,
metalliferous and petroleum resources.

A geologist engaged on exploration is often in the vanguard of any team
his organisation has assembled for a project.   He has to obtain information on a variety of topics that lie outside geology as well as obtain
a comprehensive overview of the geological parameters.   He is a jack-of-
all-trades as well as a master-of-his-own.

It is a fact that the more senior a man becomes in his profession the less
time he has to read detail.   The applied geologist is no exception; and
his information requirements are very divers.   He will seek information
from the standard geological reference sources - maps, published exploration data, and journal articles which provide new data or interpretive
techniques.   In addition he will need drilling and other sub-surface
prospecting results, and relevant production statistics.   All facets
mentioned so far are obviously associated with geology, but the applied
geologist must consider environmental constraints, glean fiscal and legal
information, and be aware of political risks.

Charter Consolidated has prepared a check list for the geologist who is
making an initial investigation which covers all these areas of a prospect:
this list contains over 200 essential points to which answers have to be
sought if the exploration and possible exploitation ventures are to
operate under the optimum conditions.   There are another 200 points on
which information will be required at an early stage in the project:
these are not essential but if the geologist can collect them he will
increase the rate at which the project is prepared for launching.

Dwelling now on the information requirements in a little more detail and considering some of the sources.

1. **Maps**

    Topographical and geological maps often have to be supplemented by aerial photography and satellite imagery. In developing parts of the world more reliance has to be placed on the photogeological information, and fortunately some of the resources such as Landsat imagery are of a very high standard.

2. **Land ownership**

    Difficulties can arise in ascertaining who owns land; but when investigating mineral ownership some virtually insoluble problems occur. Nevertheless at a very early stage in an exploration project, exploration, mineral- and mining-rights have to be established. The applied geologist is required to gather information on rights.

3. **Infrastructure**

    The accessability of a potential mineral resources extraction site is of necessity a very important factor, as are the availability of power and water services. Frequently the exploitation company will have to make access roads and/or an airstrip, and it should be established if government assistance might be available for this purpose. Infrastructure information is definitely an "essential" type.

4. **Borehole data and drilling facilities**

    There may have been drilling undertaken in the area of interest already, but whether this is available to the geologist is another matter. It is very likely that more drilling will be required and hence availability of drilling facilities needs to be known.

5. **Geological interpretation**

    A fair amount of information can be found in the literature which will aid interpretation, but the availability of raw geological, geophysical and geochemical data can be extremely variable. The applied geologist will be required to seek out what is available and recommend programmes of further investigation to obtain the additional data necessary to evaluate the deposit.

6. Background information

General information on the country in which the area(s) of interest
lie(s) can be found from various directories and periodical reviews.
Bank Reviews, national handbooks and the specialised industry directories provide varying amounts of economic and demographic details
which serve to build up a picture of the business climate in which the
venture will operate.   Stockbrokers' reviews can be a useful source of
information in a number of well established fields.

7. Government legislation

Governments are very keen to ensure any organisation exploring, and
possibly eventually exploiting, mineral resources in their country
know what the "rules and regulations" are.   Problems may arise in
translation and even the most linguistically accomplished geologist
will need help here.   The U.S. Bureau of Mines' "Mining Laws of the
World" is a very valuable reference tool.

8. Production information

Generally speaking oil and coal production figures are more abundant
than those for metalliferous deposits.   Industrial minerals production information availability varies according to the size of the
operations; those industrial minerals which are extracted by small
firms are often not well documented.   An applied geologist who is
conversant with throughput of processing plant should form an estimate
of the likely production figures for an operating quarry or mine by
observation, but this needs a lot of experience.   Another U.S. Bureau
of Mines' publication "Mineral Facts and Problems" can provide a
wealth of useful figures.   If relevant institutes or trade associations exist they can be useful sources of information, but these vary
greatly from commodity to commodity.

In many cases there is no substitute for visiting the country of interest.
The national Geological Surveys, universities, embassy commercial attaches,
banks and the internationally represented accounting houses can provide a
lot of information which will enable the geologist to obtain a "feel" for
the region.   Jointly with metallurgical and engineering colleagues the
applied geologist can piece together much information which forms the
foundation of knowledge that is essential for a successful venture.   It
is difficult to stipulate a hard and fast rule for the time required to
gather this data, but one is talking in terms of a year to establish all

the necessary contacts which will provide essential local knowledge. Rapid political changes in some countries can mean the constant up-dating of a dossier will require a considerable amount of effort.

Reference has been made already in this conference to bibliographic services, many of which are held on computer storage. These are excellent in many ways, but are biased towards academic requirements. One must also be aware that although one may retrieve an impressive set of references, obtaining the full text of some of these is not easy; and the applied geologist often wants his information quickly.

The final point to be recognised is that dependent upon the stage a project has reached information retrieval processes differ. At the start published material may only be a small proportion of the total information available much will have to be gleaned from a multitude of sources. An organisation with an efficient internal information service will be able to retrieve a certain amount from files on past, similar projects: also the information service will have built up a network of useful external contacts. When the project is underway well-managed filing services are a key to successful retrieval. When the project has terminated, or all the development stages leading to exploitation of the resource have been completed, it is vitally important to prepare a summary of the information gathered, recording the sources from which it was obtained. Sometimes the geological aspects of a project are incorporated in a report that covers all disciplines involved. Whether or not the geology report is part of the overall project report or is written separately, it is important to record what was done: unfortunately this is not always carried out, and a lot of corporate expertise is dispersed to be reassembled only with a great amount of effort and "re-inventing of the wheel".

Having pointed a broad-brush picture which sets the scene for this session, we now have contributions which present international resource assessment programmes, and accounts of activities in the specific fields of geothermal energy, coal and petroleum information.

REFERENCES

UNITED STATES BUREAU OF MINES 1970/74.   Summary of Mining and Petroleum
    Laws of the World.   Washington
    Part 1   Western Hemisphere   Information Circular 8482.   1970
    Part 2   East Asia and the Pacific   "   "   8514.   1971
    Part 3   Near East and South Africa   "   "   8544.   1972
    Part 4   Africa   "   "   8610.   1974
    Part 5   Europe   "   "   8631.   1974

UNITED STATES BUREAU OF MINES 1975.   Mineral Facts and Problems.   Bi-
    centennial Edition.   Washington
    1266pp   U.S. Bur. Mines Bulletin No.667

Dr. Allen L. Clark and Ms. Jennifer L. Cook
U.S. Geological Survey
Office of International Geology
917 National Center
Reston, Virginia 22092

Summary: The ever increasing demand for energy and mineral resources
has created an equally large demand for data pertaining to these
resources. The basic data requirements are for location, quantity
and quality of a resource. Although basic data requirements are
easy to define, the acquisition of such data is difficult and at
present data is inadequate to meet the needs of international
resource analysis. The inadequacy of data can be categorized in
three main areas: availability, credibility and applicability.

Many of the problems associated with adequate resource information
are caused by the wide geographic distribution of resources. Copper
serves as a good example of the wide geographic distribution of
mineral resources with the 241 major copper mines of the world being
in 37 nations. In terms of energy resources, such as oil and gas,
the major producing countries represent virtually every continent.

Analysis of world resources can be divided into 3 major categories
of activity based on the objectives of the analysis; inventories,
resource assessments, reserve estimates.

The objective of establishing a national inventory is to locate,
describe, and quantify resources for future assessment, planning,
and exploration purposes, as well as for policy formulation and
decision-making purposes.

There are two possible objectives in a resource assessment. First,
to determine the unconstrained amount of a resource in an area; of
which no more than this amount exists and secondly, to determine the
amounts of the resource available under differing specified economic
and technological constraints.

The major objective of a resource estimate is to systematically
determine the commercial availability of a commodity, within the
limits of available working data.

The impact of inadequate data to undertake resource analysis seriously constrains conceptual resource model development, produces
poor local, regional, and world planning, and provides an inadequate
data base for development. These problems lead to inefficient utilization of the finite energy and mineral resources of the world.

Introduction

Recent prognostications on world resource sufficiency range from
a world rapidly running out of resources to a world of infinite
resources. To most scientists it seems a rational assumption that
the truth must lie somewhere between these extremes. What can be
said, however, with respect to world resources, is that there is an
ever increasing demand for resources and although economic and social
development varies from country to country, as a consequence so does
each contry's need for raw materials, there can be little doubt of
the interdependency of the world in terms of meeting the needs of
society for natural resources. The simple fact that no nation is
totally self-sufficient in all natural resources dictates that each
country has a vital need to assess its natural resource endowment and
that of the rest of the world if it is to be assured of a continuing
source of necessary commodities. These factors further dictate that
there is also a need to undertake an assessment of the world's resources so that effective planning can be undertaken to ensure an
adequate supply of natural resources. However, for such evaluations
to be of real use they must be conducted within a framework of
agreed definitions and standardized approached to resource evaluations.

Data Definitions

Two major attributes of resources are most often sought in resource
evaluations; i.e., reserves and resources. Normally a great deal of
confusion exists with respect to the definition and classification
of reserves and resources. The relationship between reserves and
resources is best shown in the McKelvey diagram (fig. 1).

For purposes of discussion and to ensure some degree of standardi-
zation in this presentation the reserve and resource definitions
(USGS Bulletin 1450-A, 1976) agreed upon for use within the U.S.
Geological Survey and U.S. Bureau of Mines of the United States
Department of the Interior are used:

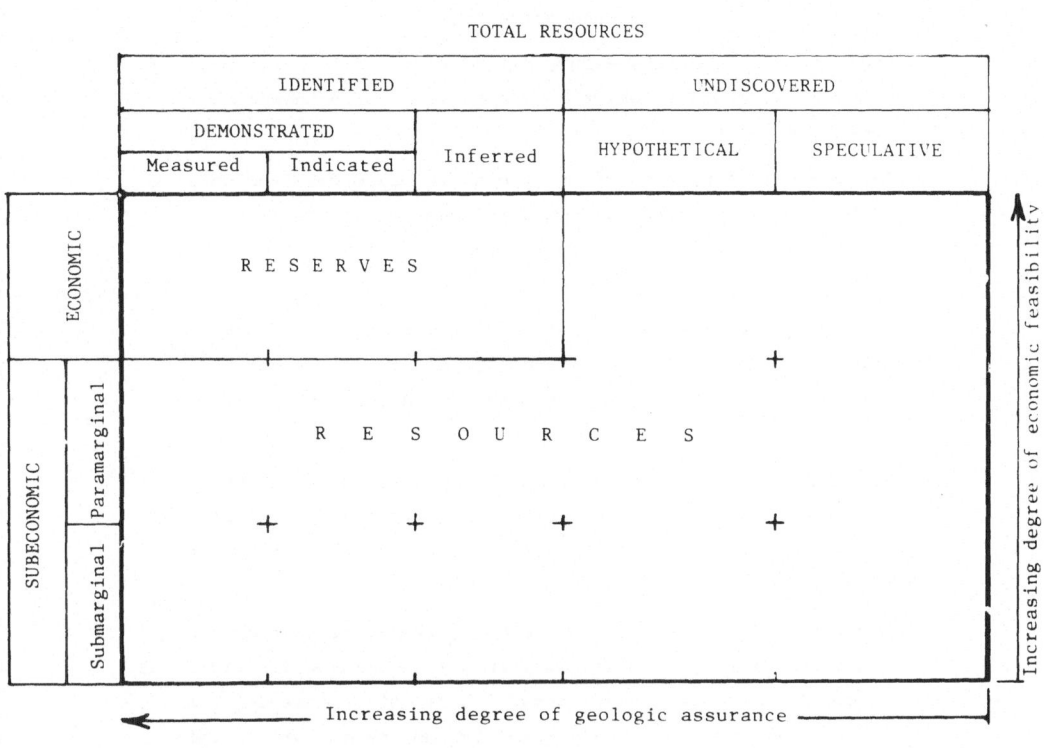

Figure 1. McKelvey Classification Diagram of Reserves and Resources

Resource.--A concentration of naturally occurring solid, liquid
or gaseous material, in or on the Earth's crust in such form
that economic extraction of a commodity is currently or potentially feasible.

Identified resources.--Specific bodies of mineral-bearing
material whose location, quality, and quantity are known from
geologic evidence supported by engineering measurements with
respect to the demonstrated category.

Undiscovered resources.--Unspecified bodies of mineral-bearing
material surmised to exist on the basis of broad geologic
knowledge and theory.

Reserve.--That portion of the Identified resource from which a
usable mineral and energy commodity can be economically and
legally extracted at the time of determination. The term ore
is used for reserves of some minerals.

Paramarginal.--The portion of Subeconomic resources that (1)
borders on being economically producible or (2) is not commercially available solely because of legal or political
circumstances.

Submarginal.--The portion of Subeconomic resources which would
require a substantially higher price (more than 1.5 times the
price at the time of determination) or a major cost-reducing
advance in technology.

Hypothetical resources.--Undiscovered resources that may
reasonably be expected to exist in a known mining district
under known geologic conditions. Exploration that confirms
their existence and reveals quantity and quality will permit
their reclassification as a Reserve or Identified Subeconomic
resource.

Speculative resources.--Undiscovered resources that may occur
either in known types of deposits in a favorable geologic
setting where no discoveries have been made, or in as yet
unknown types of deposits that remain to be recognized.
Exploration that confirms their existence and reveals quantity and quality will permit their reclassification as
Reserves or Identified-Subeconomic resources.

Careful analysis of the above definitions will clearly show why it
is so important to specifically define what aspect of the resource
one is attempting to define, since each aspect requires a unique set
of parameters in order to undertake a meaningful evaluation.

Resource Data Base

Resource data files, loosely called data bases, contain the basic
raw or disaggregated information on deposits or individual commodities.
This information is rigorously and logically organized down to the
smallest unit of information for which computer access is desired
and these individual data items are stored in computer fields, which
are individually addressable.

Although the organization of information into a computerized data file
is a tedious process, so it is also with a manual file. In the end,
in those situations where the given subject is complex, as in the
situation of mineral resources, the computerized operation offers
some unique advantages not available elsewhere. These advantages
stem from the fact that the computer file can be addressed down to
the level of the individual field, whereas the manual file only can
be addressed down to the level of the record (document) plus those
two or three fields (within the record) that have been sorted in
advance. Therefore, only the computer file can disassemble the
record into its individual parts and for most uses of data this dis-
assembly process is an initial requirement.

It is important to decide beforehand the smallest level of information that will be required because, whereas it is possible to aggregate smaller components of data into larger groups (grouped data), it is not possible to go in the reverse direction.

The content of a data base differs greatly depending on the number and scope of activities that are to be undertaken with the file. This dependency on use, to determine the structure of a data file, may be overlooked by individuals designing data files and emphasis is placed on gathering all data rather than that which is needed by the user community. The general catagories of information that normally should be organized and collected to undertake a resource inventory are:

Record Identification
Record Number
Deposit Number
Cross Index Number
Date

Name and Location
Deposit
Mining District/Area/Subdistrict
Country
State
County
Latitude
Longitude

Description of Deposit
Commodities Present
Size of Deposit (Large) (Medium) (Small)
Ore Materials
Status of Exploration/Development
Property is: (Active) (Inactive)
Workings are: (Surface) (Underground) (Both)

Analytical Data

Production/Reserves

    Annual Production and Grade

    Cummulative Production and Grade

    Reserves and Grade

    Reserves and Potential Resources and Grade

Resource Distribution

Particular care must be taken in defining the scope of the problem in terms of estimating world resources. On the surface it seems almost an insurmountable task because of the enormous number of mines and prospects in the world. A closer analysis, however, shows that 1,072 mines account for 90 percent of all mineral resource output (excluding coal); excluding the socialist block. Figures 2 and 3 show also the widespread distribution of mineral resources both geographically and by commodity.

Many of the problems associated with adequate resource information are caused by the wide geographic distribution of resources. Copper serves as a good example of the wide geographic distribution of mineral resources with the 241 major copper mines of the world being in 37 nations. In terms of energy resources, such as oil and gas, the major producing countries represent virtually every continent.

Resource Analysis

Once an inventory of a known mineral resource has been completed and a data file developed, a large number of studies can be undertaken using this inventory. Given a set of raw data, ideally computer-based, to work from, it becomes possible to aggregate these data into various groups and combinations and to produce a wide variety of end products. One important use of a resource data base is to provide the primary input to resource models which, in turn, are used to estimate the availability of minerals.

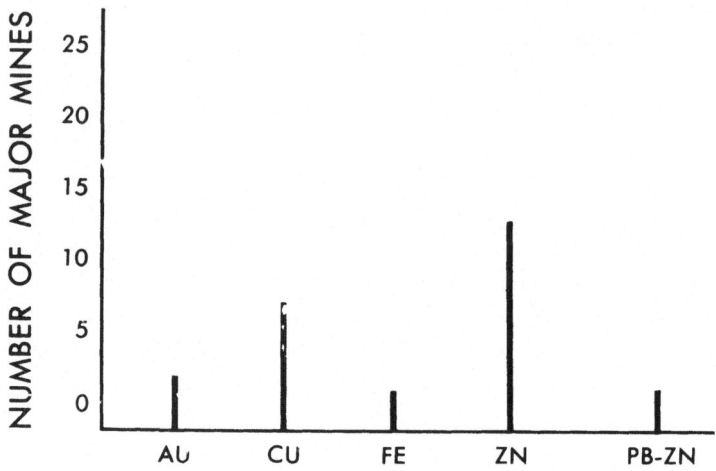

Figure 2.   Major mines by product in Peru

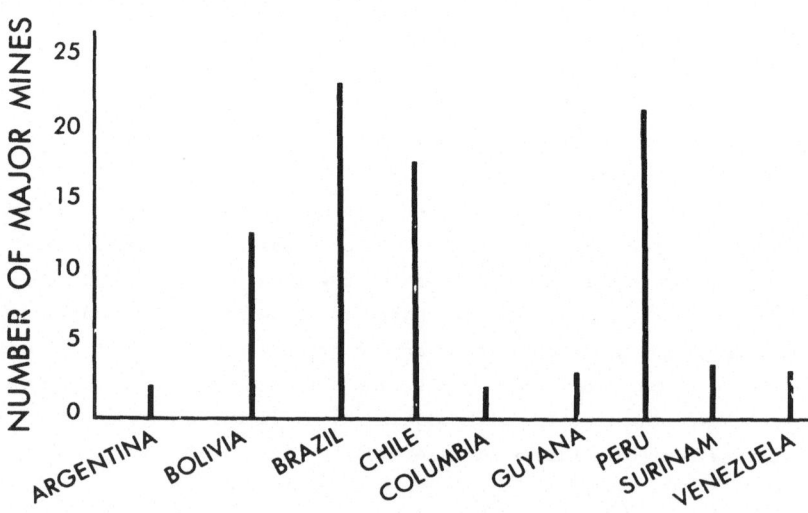

Figure 3.   Mines in South America which produce 150,000 tons ore
per year (open pit + underground)

A model is a system of postulates, data, and inferences, which serves
to predict the outcome of an event under various or changing conditions.
In describing the functions of a model, Singer (1976) states that "the
least exacting demand that can be made of any model is that it serves
as a device whereby we can predict actual physical happenings. Another
demand which could be made is that the physical happenings be in some
way relevant to man, either by allowing him to anticipate future
uncontrollable events or by demonstrating the possible consequences
of various decisions."

To date, many mineral-resource models have been deficient in meeting
these demands - because of inadequate data, insufficient knowledge of
fundamental relationships, and misunderstandings between geologists
and economists.

Previous resource estimates have been made in several different ways
and for different purposes; for example, Hubbert (1962, 1967) pre-
dicted on an aggregated basis the ultimate producible petroleum from
the conterninous United States under the assumption that technologic
and economic conditions will change at the same rate in the future
as they did in the past. The method, however, does not separately
identify that quantity of the resource which would become reserves as
a result of a radical change in price or from a technologic breakthrough in recovery technology. Moore (1966) used a similar technique
to that of Hubbert, but with far more optimistic results. By incorporating explicitly the declining efficiency in exploration, growth
of reserves through additions, and the effects of price changes,
Arps, et al (1971) developed an additional technique useful in making
resource estimates similar to those made by Hubbert and Moore.

Brinck (1971) developed a method to make highly aggregated estimates
of the availability of an element at different concentrations, tonnages, and costs. These estimates are global in scope and therefore,
so highly aggregated that they are of little use to decision makers
faced with solving specific problems in mineral resource supply. In
contrast to the global estimates made by Brinck, uranium resource

estimates were made by Bieniwoski, et al (1971) in which price, deposit
type, and recovery method were explicitly categorized. Unfortunately,
no estimates of precision are presented. Also, some may interpret
the price associated with each category as fixed, whereas these prices
may drastically change if a change in recovery technology should
occur.

Sheldon (1964) offers an example of how an extrapolation method based
upon geologic reasoning can be used to delimit favorable stratigraphic
environments for the occurrence of phosphorites. Palaeomagnetic and
palaeographic information was used in this production. The same basic
extrapolation method based upon geologic principles was used by Lowell
(1970) to estimate the number of porphyry copper deposits remaining to
be discovered in the southwestern United States, British Columbia,
Chile, and Peru. Using a similar concept, Derry (1973) estimated the
quantity of five mineral commodities remaining to be discovered in
the area north of the 60 degree latitude in Canada. Derry used the
average tonnage of each mineral commodity discovered to date per
square kilometer within each stratigraphic or tectonic rock group
to project the quantity of each of these mineral commodities remaining
to be discovered in similar stratigraphic and tectonic rock groups
occurring in northern Canada. Griffiths and Singer (1971) tabulated
the value of non-renewable natural resource production per square mile
in each state in the United States, each province in Canada and
Australia, and for New Zealand. In the United States the cumulative
production of non-renewable natural resources through 1966 averaged
250,000 dollars per square mile. Assuming that the mineral industry
is in a stage of advanced maturity, Griffiths and Singer argue that
this average value can form a reasonable first approximation of the
expected unit value of resources to be discovered in other regions
of the world. Comparisons were then made between the expected value
and the cumulative production to date in Australia, Canada, and New
Zealand. This analysis was also extended to cover the unit value
of mineral resource production of the Archean Shield area of Canada.

The concept of basing mineral resource appraisals upon geological extrapolations was extended by employing subjective probabilistic estimates of undiscovered mineral potential in northern British Columbia and the Yukon by Harris, et al (1970). Geological opinion of the occurrences of undiscovered mineral deposits of various tonnages and grades in each 36 cells was estimated through questionnairs. The purpose of the study was to provide an important input for the assessment of the placement of new rail facilities in this region.

Hendricks (1965) also used the technique of geologic extrapolation to estimate the ultimate petroleum in place in the total volume of sedimentary rocks in the United States. This extrapolation is based upon the following assumptions:

1. The historical pattern of growth in petroleum reserves through discovery and additions can be projected into the future to develop an ultimate petroleum resource estimate.

2. The remaining unexplored rocks have, on the average, half the potential for containing petroleum as the rocks previously explored.

3. Only about one-seventh of the favorable rocks in the United States have been explored.

4. Five-eights of the ultimate oil in place will be discovered. This factor is based upon an assumption that an additional two million feet of wildcat drilling will be economically justified, and that geology and geophysics will guide this drilling into the better half of the favorable unexplored rocks, and that these rocks will have an incidence of petroleum three-fourths as great as the previously explored rocks.

117

5. Of the 1,000 billion barrels petroleum projected to exist
    40 percent (400 billion barrels) will ultimately be produced.

These assumptions lead to a highly aggregated estimate which is of
limited use of solving the problems facing decision makers. Mallory
(1973) has extended Hendrick's original approach by disaggregating
the total sedimentary volume of the United States according to major
stratigraphic units. Secondly, Mallory constructed a physical ex-
haustion factor by tabulating the number of wildcat wells which have
penetrated each of the major stratigraphic units, and by assuming
that each wildcat well exhausts a specific volume of rock. Litho-
facies maps were employed to isolate the prospective from nonprospective
rock volumes. Petroleum resources estimates were then made of both the
probable and possible plus speculative by assuming that the potential
of the unexplored rock volume in each major stratigraphic unit is
equal to the productivity of the explored (exhausted) volume of rock
in each unit. This disaggregated approach offered by Mallory is a
significant improvement because it can, with the inclusion of economics and technological data, be extended to become the type of
resource appraisal useful to decision makers.

Allais (1957) relying in part on the work of Nolan (1950) built a
decision model for appraising the economic prospects of exploring a
large part of the Algerian Sahara. His resource estimate consisted
of the statistical distributions of the expected number of metal
deposits and of their distributions according to value. The distributions were estimated from known mineral occurrences in well explored
areas. Although the methodology employed by Allais and extended by
Slichter, et al (1962) was ideally suited to the type of problem
Allais was trying to solve; it is too highly aggregated to be useful in
solving most of the problems facing decision makers today and in the
near future.

The total value of metal production plus reserves was estimated for cells in Utah based upon the relationships among the geological variables and total metal value of cells in New Mexico and Arizona by Harris (1966).

A resource model involves geology, economics, and technology and is intended primarily as a tool for making resource estimates for use in the formulation of policy decisions relating to our minerals posture. The ideal objectives are to provide resource estimates that will provide an indication of how much of each commodity will be available under all conceivable future conditions.

It may seem that these objectives are open-ended and not answerable and to some extent this is true. Not all future supply problems can be anticipated and, as a result, a broad base of resource information is needed to even attempt an answer. Variations of certain policy questions have, however, tended to repeat themselves and we may expect them to recur in the future. Some of these questions, as they relate to resource estimates, are: What would be the change in reserves with a change in price of the commodity? Should some particular area be excluded from mineral development? How much additional material would be available as a result of technological breakthrough in mining methods? Is it worthwhile to encourage domestic exploration for some mineral? Should there be a mineral stockpile?

As seen from the questions, geologic, economic, technological, and political factors determine whether a mineral will be produced from a property. However, geologic factors are the independent or first-order variables in relation to the others and, therefore, only the geologic factors normally need be recorded and permanently stored in a data file. In particular, the raw data used in resource models to make resource estimates are those geologic factors which describe

the quality and quantity of the resources available and which may
have a bearing on the technology and economics of a mining operation.
Examples of such geologic factors are: grade and tonnage estimates;
mineralogy and distribution of the ore; deposit type; size, depth
and shape of the ore body; geologic structure; and geographical
location.

Factors specific to economics and technology, such as price and processing methods, need not be stored permanently in the raw data because
they are subject to change over short periods of time. Instead, they
are values which are assigned externally when it is desired to estimate
the effects of changes in economics and technology.

Thus far, resource models and resource estimates have been discussed
as they apply to known mineral deposits. Deposits that have not yet
been found also will affect the long-term supply picture and it would
be of great value if improved methods were available to predict the
amount of ore that will be available in deposits not yet found.

Single rare events, such as the occurrence of unfound mineral deposits,
are unpredictable without extensive drilling. However, the statistical
averages of large numbers of them are highly regular and predictable,
and this indirect approach can be used to estimate the magnitude of those
resources not yet found. The elements of this method are summarized
as follows:

1. statistical analysis of grades and tonnages of known mineral
   deposits is used to make models for undiscovered deposits;

2. by geologic extrapolation based upon knowledge of known
   deposit types, the number of unfound deposits of the same
   type is estimated;

3. using frequency distributions of grades and tonnages of
   known deposit types, together with the estimated number
   of unfound deposits of (2), resources of the unfound
   deposits are estimated.

Resource estimates obtained from such statistical models must be
tempered by the quality of the geological extrapolation by which the
number of unfound deposits is obtained, as well as other extrapolated
conditions such as depth, general location, ground water, lands closed
to development, and the unlikelihood of finding all of the unfound
deposits. Estimates of the precision of each resource estimate
should be stated clearly so that the uncertainty of the resource
estimates can be considered in making policy decisions.

The Office of Resource Analysis is developing a resource model that
can be applied to the formulation of many resource estimates and
which is directed towards a decision-making end product. The generalized model is presented in figure 4. Although figure 4 is simplified
in terms of the amount of data required and the effort involved in
developing such a model, it does show the basic framework within
which a resource estimate can be produced. Figure 4 is an integration
of three separate models: an occurrence model, a search model, and
an economics model.

The occurrence model gives an estimate of a resource in terms of
geologic availability. It utilizes the data from the resources data
file of the known deposits or from subjective estimates and provides
by aggregation the estimated total resources for deposits of the same
type.

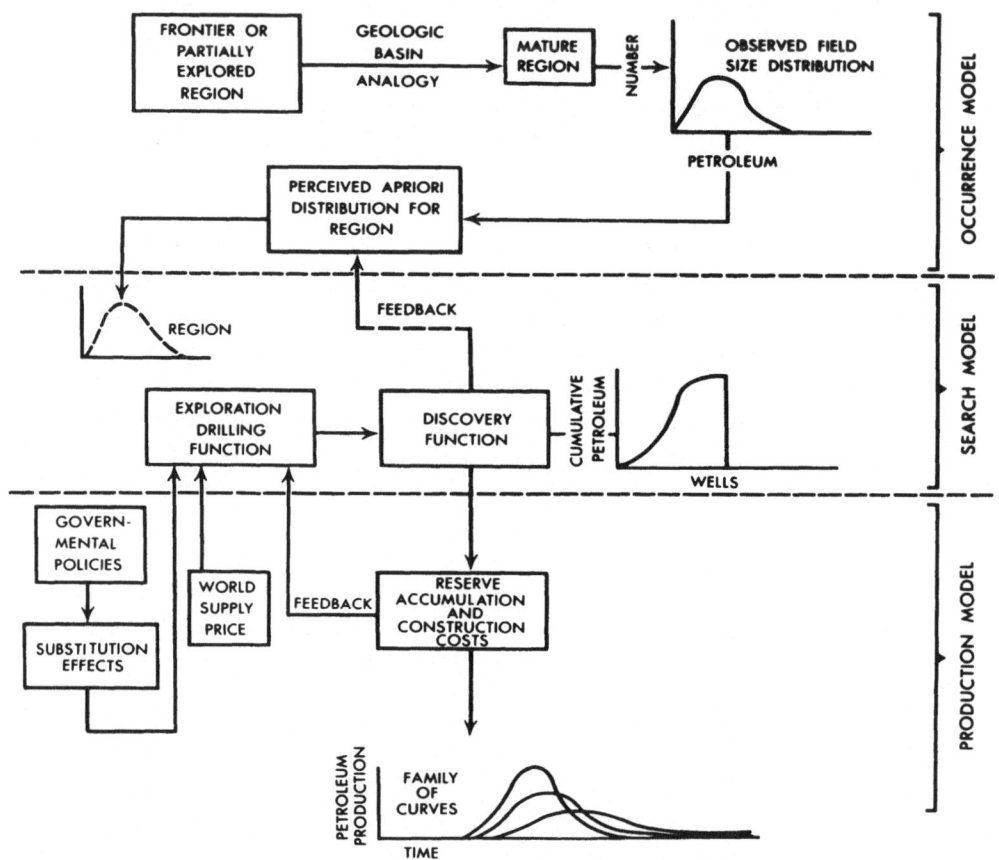

Figure 4.   Clark-Drew Conceptual Model of Petroleum Supply System

The search model provides predictive estimates for those unfound deposits which ultimately will be discovered. The search model estimates the resources likely to be found, given the resources of the occurrence model, and the technology and costs associated with exploration and development. Aggregation of the data furnished by the search model gives, therefore, the resource availability at a given price and for a given technology.

Finally, it is necessary to provide an econometric model in order to take into account the price schedules and other highly changeable economic factors relating to production, supply, and demand. This aspect of the general model is outside the scope of the U.S. Geological Survey but must be integrated with the data produced by the search and occurrence models in order to provide the short and long-term supply forecasts needed in the formulation of minerals policy.

Conclusions

The ever-increasing importance of natural resources to the world community has resulted in a demand for more and better resource data and for estimates of ultimate resource potential of individual nations and for the world in the aggregate. The most critical of the needs for   resource data and analysis are: (a) the increasing demands on the world's resources by both the developed and developing nations, (b) the need for more accurate data for short-, intermediate-, and long-range planning, (c) the need for assessing national endowment in order to effectively allocate other scarce commodities such as people and capital, and (d) the need to ascertain the best trade offs; in terms of world food demands, environmental degradation, rational growth programs and multiple land use within the nations of the world.

In order to reach the goals of adequate resource data and resource
analysis it is of paramount importance that a standardized inventory
of the world's mineral and energy resources be undertaken and completed. Once such an inventory has been completed and the basic
disaggregated data are compiled it will become possible to undertake
resource analysis programs which can effectively provide the studies
necessary to answer the questions outlined above.

The major thrust, however, of a total resource assessment should not
be overlooked because in the final analysis the total mineral and
energy resource of the world represent a finite constraint on man's
environment, way of life, and ultimately man himself.

REFERENCES

Allais, M., 1957, Method of Appraising Economic Prospects of Mining Exploration Over Large Territories; Algerian Sahara Case Study Mgt. Sci., Vol. 4, No. 4, pp. 235-247.

Arps, J. J., Mortada, M., and Smith, A. E., 1970, Relationship Between Proved Reserves and Exploratory Effort: Soc. Petroleum Engineers, SPE paper 2905.

Bieniewski, C. L., Persse, F. H., and Brauch, E. F., 1971, Availability of Uranium at Various Prices for Resources in the United States: U.S. Bur. of Mines Inf. Circ. 8501, 92 p.

Brinck, J. W., 1972, Prediction of Mineral Resources and Long-Term Price Trends in the Non-Ferrous Metal Mining Industry, Section 4, pp. 3-15, Mineral Deposits, 24th Session of the International Geological Congress, Montreal 1972.

Derry, D., 1973, Potential Ore Reserves-An Experimental Approach, Western Miner, October, 1973.

Griffiths, J. D., and Singer, D. A., 1971, Unit Regional Value of Non-renewable Natural Resources as a Measure of Potential for Development of Large Regions, Special Publication No. 3, Geological Society of Australia, pp. 227-238.

Harris, D. P., et al, 1966, A Probability Model of Mineral Weather, Transactions A.I.M.E., Society of Mining Engineers, pp. 199-216.

_____, Freyman, A. J., and Barry, G. S., 1970, The Methodology Employed to Estimate Potential Mineral Supply of the Canadian Northwest--An Analysis Based Upon Geologic Opinion and Systems Simulation: Mineral Information Bull. MR105, Department of Energy, Mines and Resources, Queen's Printer, Ottawa, Canada, 56 p.

Hendricks, T. A., 1965, Resources of Oil, Gas, and Natural Gas Liquids in the United States and the World: U.S. Geol. Survey Circ. 522, 20 p.

Hubbert, M. K., 1962, Energy Resources: Nat'l. Acad. Sci. - Nat'l. Research Council Pub. 1000-D, 141 p.

_____, 1967, Degree of Advancement of Petroleum Exploration in United States: Am. Assoc. Petroleum Geologists Bull., Vol. 51, No. 11, pp. 2207-2227.

Lowell, J. D., 1970, Copper Resources in 1970: Mining Eng. Vol. 22, No. 4, pp. 67-73.

Mallory, W. W., 1973, Accelerated National Oil and Gas Resource Evaluation, U.S. Geol. Survey Unpub. Working Paper, May 1973, 67 p.

McKelvey, V. E., 1972, Mineral Resources Estimates and Public Policy: Am. Scientist, Vol. 60, No. 1, p. 32-40.

Moore, C. L., 1966, Projections of U.S. Petroleum Supply to 1980, Washington, D. C.: U.S. Dept. Interior, Office of Oil and Gas.

Nolan, T. B., 1950, The Search for New Mining Districts, Econ. Geol., Vol. 45, pp. 601-608.

Sheldon, R. P., 1964, Paleolatitudinic and Paleogeographic Distribution of Phosphorite: U.S. Geol. Survey Prof. Paper 501-C, p. C106-C113.

Singer, D. A., 1976, Mineral Resource Models and the Alaskan Mineral Resource Assessment Program, in Non-fuels Minerals Models: a state of the arts review, Resources for the Future: John Hopkins Press, 419 p.

Slichter, L. B., Dixon, D. J., and Myer, G. H., 1962, Statistics as a Guide to Prospecting, Sym. on Mathematical and Computer Appl. in Mining and Expl.: College of Mines, Univ. of Arizona, p. FI1-FI27.

U.S. Geol. Survey - U.S. Bur. Mines, 1976, Principals of the Mineral Resource classification system of the U.S. Geological Survey and U.S. Bur. of Mines, U.S. Geol. Survey Bull. 1450-A.

INTERNATIONAL DATA EXCHANGE

FOR

GEOTHERMAL ENERGY POWER PRODUCTION

Sidney L. Phillips

Lawrence Berkeley Laboratory

University of California

Berkeley, CA 94720

Summary: During the past five years great strides have been made in the
development of geothermal energy resources for electrical power production.
However, the same time has seen an enormous growth in publications dealing
with geothermal energy, thereby raising the question of economic data
handling and dissemination. An approach to this problem is the establishment
of an international data exchange cooperative program with the idea to avoid
unnecessary and expensive duplication of research and development effort.

Introduction

The international effort to develop and utilize geothermal energy has resulted in an enormous growth of information. However, the needed data is widely scattered, difficult to access, and largely unevaluated. Thus, an important task is to collect, evaluate and disseminate geothermal data in a timely manner, thereby avoiding unnecessary and expensive duplication of research effort.

The National Geothermal Information Resource (GRID) of the Lawrence Berkeley Laboratory is sponsored by the U.S. Department of Energy (DOE) to provide critically evaluated information for the development and utilization of geothermal energy. Included are both site specific and basic information related to resource evaluation, electrical and direct utilization, environmental aspects, and the thermodynamic and transport properties of aqueous electrolytes.

The GRID project is involved in cooperative agreements for the interchange of information with other organizations. There are currently three U.S. data centers working to implement the collection and exchange of information on geothermal energy research and production: the DOE Technical Information Center (TIC), Oak Ridge, the GEOTHERM database of the U.S. Geological Survey in Menlo Park, and the GRID project. The data systems of TIC, GEOTHERM and GRID are coordinated for data collection and dissemination, with GRID serving as a clearinghouse having access to files from all geothermal databases including both numerical and bibliographic data. GRID interfaces with DOE/TIC for bibliographic information and with GEOTHERM for certain site-dependent numerical data. Data exchange at the international

level is mainly with the Geothermal Data Bank CNUCE, Pisa, Italy via
DOE.

It is not possible to discuss all facets of international data exchange for geothermal energy power production. This report is limited
to the following aspects of information exchange at the international
level: description of the database; data formats; the NATO-CCMS pilot
study; and a section on recommendation. for information exchange based on
a one-year pilot study under NATO-CCMS.

Database Management

The Berkeley Data Base Management System (BDMS) is used for creating,
maintaining and accessing both bibliographic and numerical data. Bibliographic
records are readily retrieved from computer files using BDMS by specifying
either one parameter such as the geothermal site, or a combination of parameters such as the geothermal site, the date and designated data measurement.
Standards for interchange of bibliographic data are patterned after that of
the International Atomic Agency's International Nuclear Information System
(INIS). Utilization of the INIS format ensures compatibility with other
INIS styled computer centers, thereby promoting the active interchange of
data with other groups [Herr(1977)]. Data is available in the form of
bibliographic compilations, numerical tables, or graphical displays disposed
to paper, film or magnetic tape.

The GRID documentation system (GEODOC) contains the descriptive
cataloging and indexing information for coding records. Each record
contains the descriptive cataloging, abstracting, and indexing information
corresponding to a single document; the information within a given record

is subdivided into data elements, some of which are indexing keys. Table I lists the definitions of typical data elements which may appear in a GEODOC record. Some data elements (e.g., author's name) can occur repeatedly within one record; an "m" in the third column of Table I indicates that such multiple occurrences are allowed. The tag used to label the data elements within a record is shown in the left hand column of Table I.

The data elements bear certain hierarchical relationships to each other; the structure is indicated in Table I by indenting the tag names of sub-ordinate data elements and placing them after their parents. Data elements are input to the system in any order except that subordinate data elements must follow the occurrence of their parent with which they are associated.

Information Packaging and Dissemination

Information packaging is providing the data in a form that will best fit the needs of the user of the database. Output therefore takes a variety of forms. Typical information packaging for geothermal data is shown in the following table:

Information Packaging

Typical User	Data Type	Format Example
Scientist; Engineer	Basic Scientific	Correlation Equation
Data Evaluator; Program Manager	Site Dependent	Tables of Numerical Values
Engineer, Program Manager	Critical Survey	Report with Recommendations
Evaluator	Annotated Bibliography	References and Indexes

Typical
Table I.   GEODOC Data Elements

LBL Tag	INIS Tag	m*	n*	Data Element Definition
SC	008			document short code: unique identifier for document
TY				type of document/bibliographic levels/literary indicator
DES-CAT		m	n	delineates information for one bibliographic level
BL	009			bibliographic level indicator
PT	200			primary title (translated into English if necessary)
PS	201			primary subtitle (translated into English if necessary)
TA	620			title augmentation
L	600			language (for non-English document)
OT	230			original title (non-English) or journal/ series title
OS	231			original subtitle (non-English) or journal/ series subtitle
ED	250			edition
CODEN				journal CODEN
AUTHORS		m	n	delineates author - affiliation group
AU	100	m		author's name
AN	100			author note (ed., comp., eds., comps.,)
AA	100	m		author's affiliation
AC	700			affiliation code
CE	110	m		corporate entry
CC	710			corporate code
DG	111			academic degree
SPO		m		sponsor
SPC				sponsor code
SCN		m		sponsor contract number

An example of packaging is shown by comparing Figures 1 and 2. The format shown in Figure 1 is a typical output from BDMS which reflects the hierarchical structure shown in part in Table 1. We have developed a report generator which reformats the bibliographic reference in Figure 1 to a more familiar form shown in Figure 2.

NATO-CCMS Geothermal Information Exchange Pilot Study

The NATO/CCMS Information Pilot Study was established to encourage cooperative exchanges of data among participant countries. In this summary of the pilot study, the following topics are discussed: (1) objectives of the program; (2) the computerized geothermal data bases involved in the exchange program; (3) results of the CCMS phase of the program; (4) current status of geothermal data exchange with traceability to CCMS; and (5) recommendations for future work on international cooperation in information exchange.

The main goal of the program was a one-year test of linked data centers designed to facilitate the exchange of new geothermal information at the international level. Three linked principal world centers were originally suggested, as described in the Summary Record of the First Geothermal Implementation Conference [Phillips (1978)].

> Three equivalent regional information centers should be established to promote international exchange of information and data concerning the utilization of geothermal resources. Suggested locations for the centers are Italy, New Zealand, and the United States. The centers would be linked in the sense that all information generated in one center would be transmitted to the other centers and each would have a complete data file of all available information.

and

> Individual requests for specific information could be made to any of the regional centers. Each center would pro-

```
RECORD 173
SC = PETERS 74;
TY = J/AS;
DES-CAT.1;
  BL = A;
  PT = CIVIL ENGINEERING FEATURES OF A GEOTHERMAL POWER PLANT;
  AUTHORS;
    AU = PETERS, S.;
    AA = PACIFIC GAS AND ELECTRIC CO., SAN FRANCISCO, CALIF. (USA);
DES-CAT.2;
  BL = S;
  OT = AM. SOC. CIVIL ENG., J. POWER DIV.;
  PUD = 1974;
  COL = V. 100 (P02), P. 157-173;
INDEX;
  CQ = EXPLORATION/EVALUATION;
  DE.1 = GEYSERS GEOTHERMAL FIELD;
  DE.2 = CALIFORNIA;
  DE.3 = SEISMICITY;
  DE.4 = SITE SELECTION;
  DE.5 = MAPS;
  DE.6 = DIAGRAMS;
  DE.7 = TABLES;
  DE.8 = GEOLOGIC SETTING;
  DE.9 = FORECASTING;
  DE.10 = GRAVITY SURVEYS;
  DE.11 = STRUCTURAL MODELS;
  DE.12 = VAPOR-DOMINATED SYSTEMS;
  DE.13 = MICROEARTHQUAKES;
  DE.14 = ENVIRONMENTAL EFFECTS;
  DE.15 = GOVERNMENT REGULATIONS;
  DE.16 = TRANSFER PIPES;
  DE.17 = POWER PLANTS;
  DE.18 = SAFETY;
  DE.19 = COOLING TOWERS;
  DE.20 = CAPITAL;
  DE.21 = ECONOMICS;
  DE.22 = GEOTHERMAL WELLS;
  DE.23 = POWER GENERATION;
  DE.24 = PHOTOGRAPHS;
CONTROL;
  DCSO = COPY ON FILE;
```

Fig. 1.

TITLE- HECDOR-A HEAT EXCHANGER COST AND DESIGN
      OPTIMIZATION ROUTINE.

AUTHOR- TURNER, S.E.;MADSEN, W.W. [IDAHO NATIONAL
      ENGINEERING LAB., IDAHO FALLS (USA)].

REFERENCE- HECDOR-A HEAT EXCHANGER COST AND DESIGN
      OPTIMIZATION ROUTINE. TREE-1112, EG AND G
      IDAHO, INC., IDAHO FALLS, IDAHO, 1977, 142 P..

DESCRIPTORS- THEORETICAL TREATMENTS; COMPUTER
      CALCULATIONS; BINARY CYCLE; POWER GENERATION;
      HEAT EXCHANGERS; COMPUTER CODES; EVALUATION;
      IDAHO; TABLES; HEAT TRANSFER; U FACTOR.

                    172

                                        FRANK 75
                        EXPLORATION/LAND-USE FACTORS

TITLE- RECURRENT GEOTHERMALLY INDUCED DEBRIS
      AVALANCHES ON BOULDER GLACIER, MOUNT BAKER,
      WASHINGTON.

AUTHOR- FRANK, D.;POST, A. [GEOLOGICAL SURVEY,
      TACOMA, WASH. (USA)].

      FRIEDMAN, J.D. [GEOLOGICAL SURVEY, DENVER,
      COLO. (USA)].

REFERENCE- J. RES. U. S. GEOL. SURV., V. 3 (1), P.
      77-87(1975).

DESCRIPTORS- WASHINGTON; MT. BAKER; SURFACE
      MANIFESTATIONS; HYDROTHERMAL ALTERATION;
      AVALANCHES; GEOLOGIC SETTING; PHOTOGRAPHS;
      TABLES; TEMPERATURE MEASUREMENTS; DIAGRAMS;
      INFRARED SURVEYS; SAFETY.

                    173

                                        PETERS 74
                        EXPLORATION/EVALUATION

TITLE- CIVIL ENGINEERING FEATURES OF A GEOTHERMAL
      POWER PLANT.

AUTHOR- PETERS, S. [PACIFIC GAS AND ELECTRIC CO.,
      SAN FRANCISCO, CALIF. (USA)].

REFERENCE- AM. SOC. CIVIL ENG., J. POWER DIV., V.
      100 (P02), P. 157-173(1974).
                    Fig. 2.

                    134

DESCRIPTORS- GEYSERS GEOTHERMAL FIELD; CALIFORNIA;
        SEISMICITY; SITE SELECTION; MAPS; DIAGRAMS;
        TABLES; GEOLOGIC SETTING; FORECASTING; GRAVITY
        SURVEYS; STRUCTURAL MODELS; VAPOR-DOMINATED
        SYSTEMS; MICROEARTHQUAKES; ENVIRONMENTAL
        EFFECTS; GOVERNMENT REGULATIONS; TRANSFER
        PIPES; POWER PLANTS; SAFETY; COOLING TOWERS;
        CAPITAL; ECONOMICS; GEOTHERMAL WELLS; POWER
        GENERATION; PHOTOGRAPHS.

174

SMITH 77B
EXPLORATION/DRILLING

TITLE- SUMMARY OF 1976 GEOTHERMAL DRILLING-WESTERN
        UNITED STATES.

AUTHOR- SMITH, J.L.;ISSELHARDT, C.F.;MATLICK, J.S.
        [REPUBLIC GEOTHERMAL, INC., SANTA FE SPRINGS,
        CALIF. (USA)].

REFERENCE- GEOTHERM. ENERGY MAG., V. 5 (5), P.
        8-17(1977).

DESCRIPTORS- DRILLING; CALIFORNIA; IDAHO; NEVADA;
        OREGON; UTAH; WESTMORELAND GEOTHERMAL FIELD;
        GEOTHERMAL WELLS; FLOW RATES; TEMPERATURE
        MEASUREMENTS; DEPTHS.

175

KUNZE 74B
EXPLORATION/EVALUATION

TITLE- IDAHO GEOTHERMAL R AND D PROJECT REPORT FOR
        PERIOD JULY 16, 1974--SEPTEMBER 30,1974.

AUTHOR- KUNZE, J.F.;MILLER, L.G.;WHITBECK,
        J.F.;RICHARDSON, A.S.;LEWIS, B.D.;MARTIN,
        L.F.;NEITZEL, J.W.;SPENCER, S.G.;STOKER, R.C.
        [AEROJET NUCLEAR CO., IDAHO FALLS, IDAHO (USA)].

REFERENCE- IDAHO GEOTHERMAL R AND D PROJECT REPORT
        FOR PERIOD JULY 16, 1974--SEPTEMBER 30,1974.
        ANCR--1190, AEROJET NUCLEAR CO., IDAHO FALLS,
        IDAHO, 1974, 12 P..

DESCRIPTORS- MAPS; DRILLING; HEAT EXCHANGERS;
        TEMPERATURE MEASUREMENTS; IDAHO; GEOLOGIC
        SETTING; RAFT RIVER KGRA; ENTHALPY; DESIGN.

Fig. 2 (cont.)

vide information in appropriate tabular, graphic or computer-readable formats.

Final implementation of the information pilot study involved computerized data centers in Italy and the United States, with active participation by New Zealand in development of formats and in providing data.

In summary, the Information Exchange Pilot Study was a one-year study, using two linked regional data centers, to develop common formats and utilize computer methods for the prompt exchange and dissemination of new information and data related to geothermal energy. The information exchange depended on computer assistance to provide the most rapid, efficient and economical means of handling large quantities of data.

Implementation of the Information Exchange Program

Implementation of the program involved the following participating data centers: (1) Centro Nazionale Universitario di Calccolo Eletronica (CNUCE), Pisa, Italy; (2) The National Geothermal Information Resource (GRID), Berkeley, California, and (3) GEOTHERM, Reston, Virginia (now Menlo Park, California).

The idea was to include both bibliographic and numerical data in the exchange on a trial basis to assess the time, cost, and usefulness of the work,   Objectives and data collection for the one-year pilot study included the following:

1. Bibliographic data compiled by GRID using CDC machines and the Berkeley Data Base Management System (BDMS) on the following aspects of development and use of geothermal energy, including information from other fields with relevance to geothermal

energy: (1) Subsidence; (2) Hydrogen Sulfide; (3) Geothermal Resources; and (4) Non-Electrical Applications. Indexed and annotated bibliographic listings were made available either as computer print-outs or magnetic tapes.

2. Tape of site-dependent numerical data by GEOTHERM using IBM machines and the GIPSY data management system to include the following subject areas: Geothermal Field/Area; Chemical Analysis; Geothermal Well/Drillhole.

3. Development of formats using internationally accepted standards by CNUCE, GEOTHERM and GRID.

4. Transmittal of computer tapes and other data to Pisa from Berkeley.

Conclusion

The results of the CCMS pilot study and the follow-up work demonstrate the effectiveness of the pilot study concept as applied to the exchange of geothermal information on a worldwide basis, and the advantages of computerized information systems for this kind of operation. The success of the work requires the cooperation and coordination of many agencies and laboratories in each participating nation. In the U.S., for example, the agencies involved in coordinating the project were the Department of Energy, Department of State, Environmental Protection Agency, U.S. Geological Survey, and Lawrence Berkeley Laboratory of the University of California.

Recommendations for Future Work on International Information Exchange

The primary objective of the CCMS Pilot Study on data exchange was to create an international geothermal resource data base, a pool of information from which all countries may draw. The difficulty does not lie with designing the data system but rather with the mechanics of securing and coding the information. Most participants are eager to contribute to the file but balk at the tedious and sometimes formidable task of coding forms for the computer. This is understandable because such coding could create a drain on manpower and funds. Future work in data exchange must face the reality that responsibility for coding lies with the data center.

Recommendations for this and other future work include the following:

1. Participating countries should collect copies of data (e.g., internal reports, manually logged data) for transmittal to the designated computer centers. This is especially important for data that is not widely circulated and may be either inaccessible or difficult to obtain. This data is important for evaluation and calculation of energy parameters and should be included in the information exchange.

2. The data centers responsible for maintaining the computer tapes should fill out the input format forms. Participating countries would be required only to provide the necessary copies of reports and other data. The computer centers should therefore make provisions to add needed staff to code the information.

3. The time required to exchange or transmit material between participating countries needs to be shortened, and site visits by computer center staff with an agreed-on frequency (e.g. yearly interval)

to collect reports is required.

4. Each participating country has a different type of data need which should be provided by the data centers. The data centers, to provide for these specific requirements, may coordinate requests for information with other organizations. It is important that participating countries be provided the data they need in exchange for their reports.

5. The computer centers should contain two types of information: (a) data evaluated by the center; (b) data evaluated by others. While it is not possible to critically evaluate all data within reasonable time frames, users of the data should be aware of the sources of the information.

6. Priorities in the acquisition of data should be established. For example, given the choice between data in publications and data in unpublished files, it might be important to concentrate on the unpublished data first. Later, the more generally available data could be secured from libraries.

7. Transcribing the data on forms for computer input is the most difficult task. Three possibilities seem most reasonable.

    a. The data would be copied and sent to the data center for encoding.

    b. People would be sent to the countries to encode the information available there.

    c. Funds would be provided to the country so it may hire someone to do the encoding.

8. A system of responding to the participating countries should be organized. Such a task may include a newsletter and periodic retrievals from the file.

9. Provide computer expertise to those developing nations which currently lack such capability. Large quantities of data are handled most effectively by a computer medium (e.g., magnetic tape); it is therefore imperative that computer expertise be initiated by nations which currently lack this capability.

References

HERR, J.J., PHILLIPS, S.L., SCHWARTZ, S.R., TRIPPE, T.G., 1977, Standards for Multilateral and Worldwide Exchange of Geothermal Data, Mathematical Geology, 9:259-263.

PHILLIPS, S.L., SCHWARTZ, S.R., SWANSON, J.R., 1978, NATO-CCMS Geothermal Information Exchange Pilot Study, UCID-3995, Lawrence Berkeley Laboratory, University of California, Berkeley, CA 94720.

DATA BANKING IN IEA COAL RESEARCH

BRIAN J. MOSS

International Energy Agency (Coal Research)

14/15 Lower Grosvenor Place, London   SW1 0EX

Summary: The International Energy Agency is an autonomous body established in November 1974 within the framework of the Organisation for Economic Co-operation and Development (OECD) to carry out research in non-petroliferous energy fields.   That part of the IEA involved with studies of coal as an energy source is based in London, England.

Two of the five internal departments (or Services) of IEA Coal Research are heavily committed to utilising data banks and data banking systems.   The first of these, the Technical Information Service (TIS) is involved in banking bibliographic data whereas the second, the World Resources and Reserves Data Bank Service (RRS) is banking quantitative data on world coal deposits.   Both TIS and RRS have carried out design studies on available data management systems and concluded that ASSASSIN and MRDS software best met their respective requirements.

Introduction

The International Energy Agency is an autonomous body established in November 1974 (within the framework of the Organisation for Economic Co-operation and Development - OECD) to implement the International Energy Programme (IEP) adopted by nineteen of the OECD twenty four member countries.   The basic objectives of this programme are (OECD, 1977):

(i)   To take measures to meet oil supply emergencies.

(ii)   To reduce dependence on imported oil by undertaking long-term co-operative efforts on energy conservation, accelerated development of alternate energy sources and, research, development and demonstration in energy.

(iii)   To promote co-operative relations with oil producing and other oil-consuming countries (including those of the developing world).

IEA energy research, development and demonstration
activities (R & D) are directed by a working party or steering group
composed of specialists drawn from IEA Member countries. At present
there are thirteen such bodies dealing with aspects of conventional
energy R & D.   These are summarised in figure 1.

Because of its experience and long tradition of involvement
in coal technology, the UK was invited to take the lead in establishing
IEA coal research (IEA Coal Research, 1977).   Implementing agreements
were signed in Paris on November 20, 1975 for five coal research projects,
which by common assent, centered on the formation of the following:

(a)   An Experimental Fluidised Bed Combustion Plant

(b)   A Technical Information Service (TIS)

(c)   An Economic Assessment Service (EAS)

(d)   A World Coal Resources and Reserves Data Bank
        Service (RRS)

(e)   A Mining Technology Clearing House (MTCH)

Of the abovementioned, TIS and RRS are both heavily involved in data
banking, the former for constructing a computerised bibliographic service,
whereas the latter is using these techniques as a research tool in its
efforts to build up a sophisticated data bank on world coal resources and
reserves.

Technical Information Service - Background

TIS is currently funded by 12 contributing countries.   These
are Austria, Belgium, Canada, West Germany, Italy, Japan, the Netherlands,
New Zealand, Spain, Sweden, the UK and the USA.   The Service aims to
report to users within these countries on world-wide developments in
coal technology and to facilitate exchange of information between participating countries (IEA Coal Research, 1978).

The interests of the coal industry cover a wide spectrum of subjects and
disciplines.   Selected information on most of these are available from
national and international organisations disseminating information in

Fig I: INTERNATIONAL ENERGY AGENCY ENERGY RESEARCH, DEVELOPMENT AND DEMONSTRATION STRATEGY

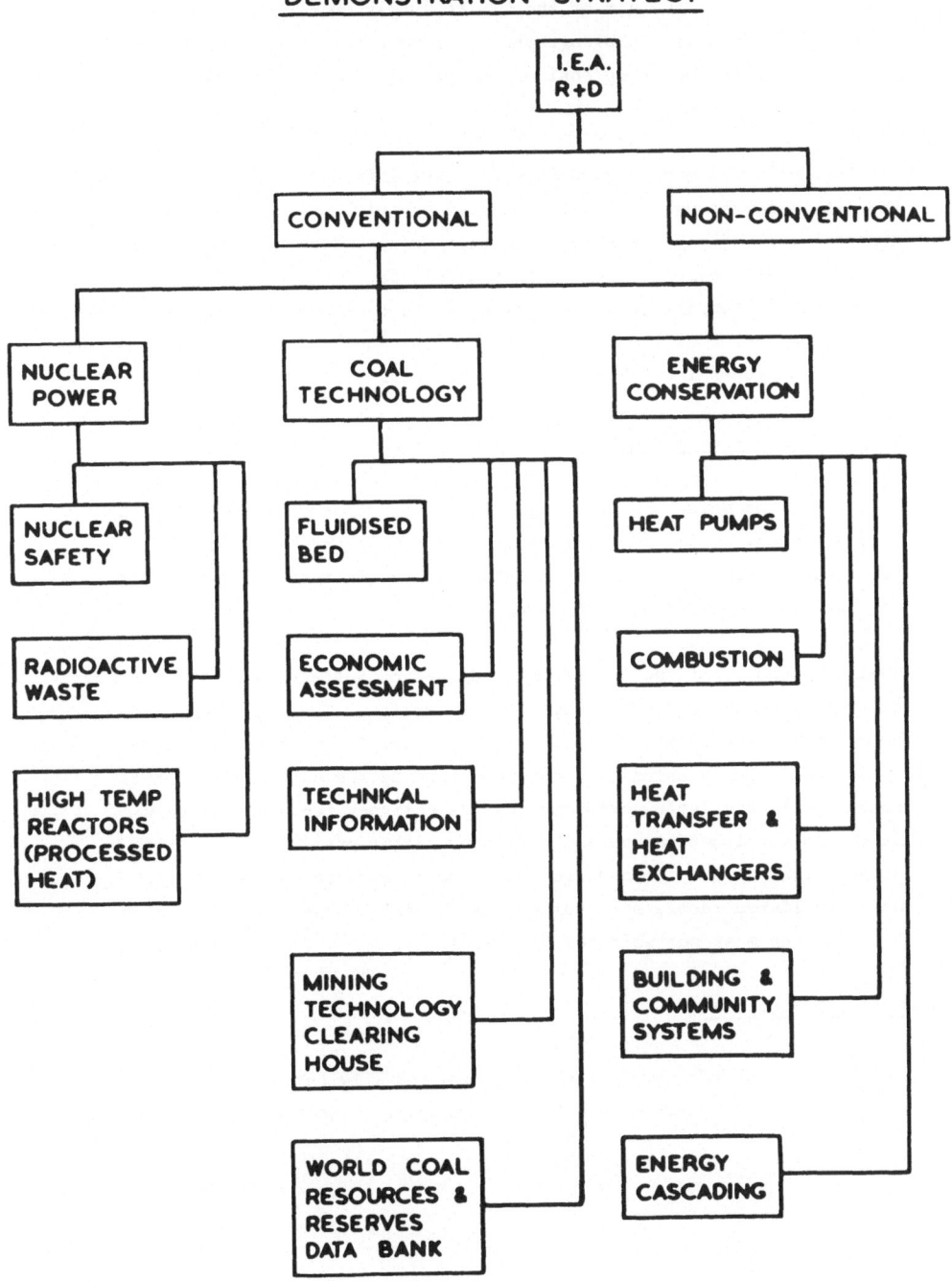

science and technology, via abstract journals and computerised data base searches but, there is no single comprehensive source to which an enquirer can resort that will give all the latest developments in coal technology. It is the aim of the TIS to supply such a source by establishing a computerised bibliographic data base containing abstracts of the literature covering coal production, technology and utilisation from all over the world.

Although users in member countries are kept informed of additions to the data base through publication of the monthly abstract journal Coal Abstracts, a major part of the Service's work is devoted to handling specific user enquiries ranging from simple fact-finding to sophisticated bibliographic compilations. A further feature offered by the Service is the production of critical technical reviews and literature summaries on subjects of general interest to coal producers and consumers.

Computer Data Base Management System Choice

The data bank facility offered by the Service is supported by a technical staff of just 6 people. This perhaps highlights the contribution that a computerised data storage and retrieval mechanism can have in the Information Sciences.

Obviously choice of the "correct" hardware and software has been largely controlled by the objectives set by the Service. The choice was made by restating these objectives in terms of what the software would need to do, and then by matching these requirements against a number of commercially available software packages designed for bibliographic date-base construction and exploitation. After preliminary investigations eight commercial packages were looked at more carefully. None were considered ideal, but choice finally fell on the ASSASSIN package as being the most suitable for TIS needs.

System Description

ASSASSIN (Clough, 1974; Clough and Kilvington 1978) is a software package developed by ICI (Agricultural Division) and first implemented

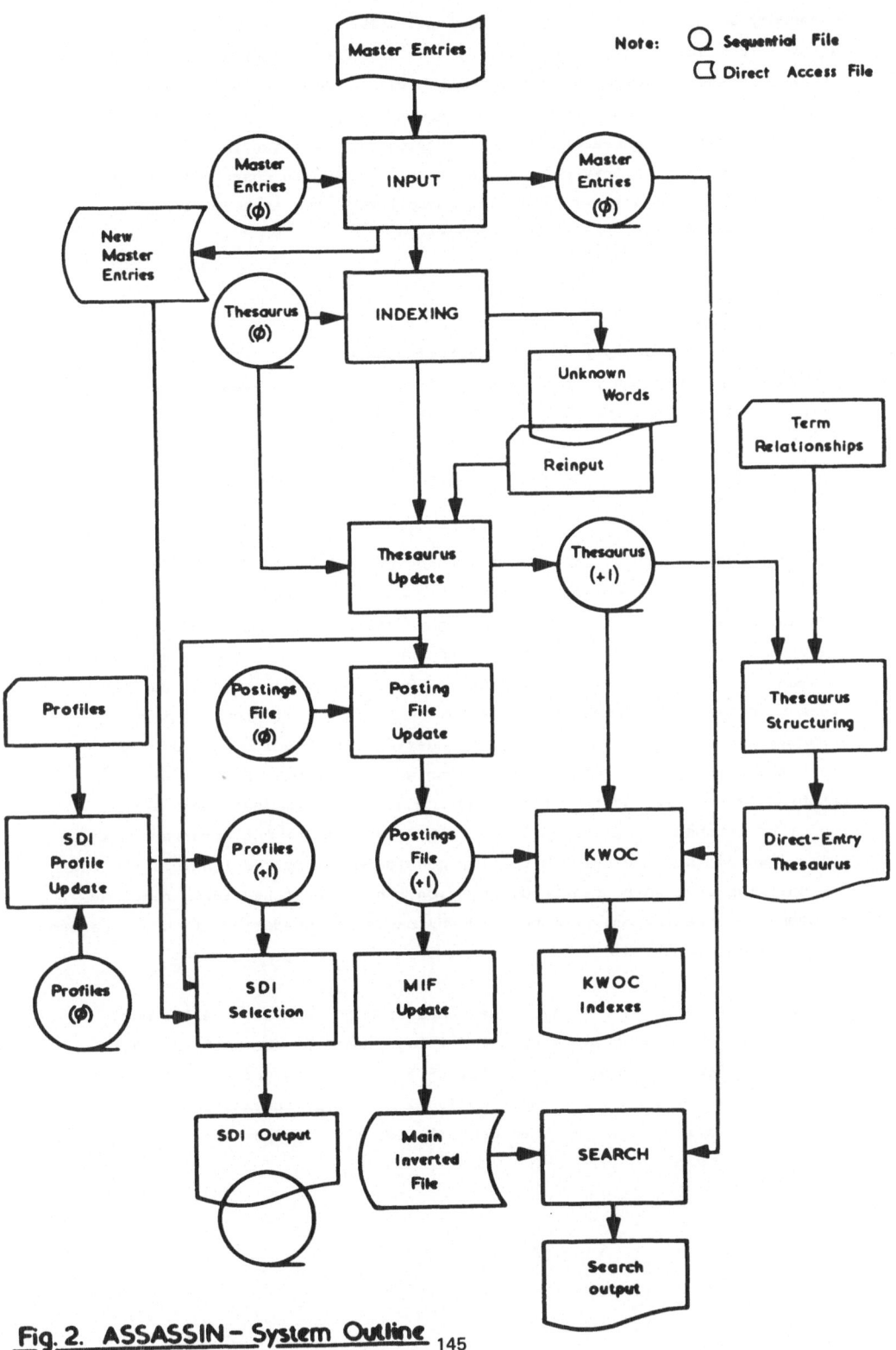

Fig. 2. ASSASSIN - System Outline 145

in January 1970. TIS are currently using the Mark V version, released
in May 1978. The package contains routines for file updating, automatic
abstract indexing, selective dissemination of information (SDI) output,
vocabulary control, the production of keyword-out-of-context (KWOC)
indexes and extensive searching. All programs are written in COBOL and
have been run on a variety of main frame computers. In essence the package
is made up of three discrete routines which handle the tasks of input,
thesaurus maintenance, and search (figure 2).

Input

        This routine processes abstracts held on magnetic tape and
punched paper tape, and produces: an SDI printout of relevant items
in the input for each customer, a machine-held thesaurus file, listings
(on request) of KWOC type indexes for the input data and an inverted
file (or index) for the input, and builds up a serial tape of all the
items entered.

Each abstract (figure 3) is given a primary key, made up of a concaten-
ation of a computer-supplied number with a user supplied two digit Source
Code and a data code, which forms a unique reference to the document within
the system. (The Source Code enables the user to divide the data base into
subject or origin groups, of which up to 99 are allowed). Input routines
break down each input abstract by title, author and text sections into
separate words, or specially assigned indexing terms, which are matched
against a computer held alphabetic Comparison Thesaurus. Successful
matching results in the automatic indexing of that item, but where words
cannot be matched (due, perhaps, to spelling errors, or the occurrence of
a 'new' word) they are printed out for consideration by the indexer. Following checking and, where required, revision, those unmatched terms required
to complete indexing are re-input and any necessary up-dating of the Comparison
Thesaurus made.

Once re-input has been completed the computer holds a file of all input
from that batch together with details of the indexing for each abstract.
From these, ASSASSIN produces selective SDI's for users and updates the
inverted files of indexing terms. Optional KWOC type indexes and a cumulated
KWOC file of all entries may then be built for later output.

date	source code		CODEN, ISSN etc	item type
[0 7,810	01	00000NZ	C,L,G,U,A, , , , ,	J,

IEA COAL RESEARCH

Title

[1 Variations in the iron content of some outcrop waters in South Durham

Reference

[2 " Colliery Guardian; 226(5); 233-234 (May 1978) =

Authors

[3 "Frost, R.C. =

Indexing terms

[4 abandoned @ shafts; iron @ compounds; sulfates; quantity @ ratio;
ground @ water; water @ pollution; coal @ mines; floods

Subject codes

[5 P-GB P-51-GB

Coal Abstracts codes

[6 12750P

PLEASE TURN OVER

Fig. 3

Abstract

[7 The iron and sulphate content of three discharges from abandoned [@] workings which have been flooded for more than 50 years are reported in an attempt to provide insight into their long term behaviour. Ferruginous outcrop waters are of two types: in one, the iron concentration should eventually stabilise at a low level; in the other, pollution will continue for an indefinite period, with seasonal variations.

Fig. 3 (continued)

Thesaurus

The Thesaurus routine acts as an 'intermediary' between the input section and the requirements of the search specification. The programs which make up this routine produce a structured 'Relationship Thesaurus' from the Comparison Thesaurus, using symbolic term identifiers (i.e. term identifier, higher term, subordinate or reltated terms) supplied at input. The Relationship Thesaurus built up by this routine contains the complete relationships between terms and families of terms.

Search

ASSASSIN can carry out searches using one of two available search strategies. The first of these enables the enquirer to batch enter search questions broken down into weighted concepts expressed using valid terms (obtainable from a printed thesaurus) and produces a listing in order of relevance. The second strategy employs nested Boolean logic allowing expressions of the type

$$((A + B) \text{ or } C \text{ or } D) + (E \text{ or } (F + (G \text{ or } H)))$$

Search output in both cases can consist of accession numbers, accession numbers with titles and/or accession numbers with full abstract printout. An interactive search facility will become available shortly.

An important additional facility offered by ASSASSIN is its interface option with printing via a computer typesetting routine which formats output to the requirements of a specified offset printer. Other advantages offered by ASSASSIN are its proven track record and, since the vendors operate a computer bureau, the opportunity of using a computer service with long handling experience and programmer support.

The bibliographic data base is being added to at the rate of between 500–600 abstracts/month, and this will increase as sources of input are widened.

Commercial Data Base Usage

The TIS managed data base is being complemented for retrospective searches by usage of four data base spinners which in turn provide access to some 85 bibliographic commercial data bases. The four data spinners in question are:-

Dialtech   –   European Space Agency, Frascati

Dialog   –   Lockhead Information System, Sunnyvale, Ca.

Orbit   –   Systems Development Corp., Santa Monica

Blaise   –   British Library

World Coal Resources and Reserves Data
Bank Service (RRS) – Background

The RRS is currently being funded by six countries: Belgium, Canada, Italy, the UK, and USA and West Germany. The Service was established to provide to its Contracting Parties information on world coal resources and reserves at differing levels of aggregation. (IEA Coal Research, 1978).

At an early stage it was realised that there are, at present, many different methods of assessing (and reporting) both coal resources and reserves. Since each country tends to use its own assessment procedures any comparisons between, and summaries based upon, data reported by differing conventions are likely to be inaccurate and misleading. Accordingly, for the purposes of unambiguous reporting, comparison and summarising of resource/reserve data, the Service must gather and analyse information from many fields of study. Broadly these include coal geology, mining, processing, transportation, marketing and utilisation together with economics and environmental control.

It was concluded that the best method of approach should take into account the following points:

(i)   The most efficient way of storing and accessing both collected (primary) and derived (secondary) data would require the use of a computerised data banking system.

(ii)   All primary data should be banked in untranslated form. This means that the translation of data based on one country's conventions to an equivalent form based on an international or other country's convention is carried out as a post-banking operation.

This ensures against the possibility of banked
primary data being degraded by translation.

(iii)   Any attempt to translate retrieved data will require
development of some form of translation procedure.
The first necessary step in the development of such
procedures is the compilation of a comprehensive
lexicon containing all pertinent definitions of and,
where possible, relationships between coal resource
and reserve classifications.

Computer Data Base System - Requirements

The points outlined above have had a fundamental bearing on
the type of data banking system adopted by the Service.

Firstly, it was realised that the 'optimum' system should be able to
build up and maintain a data base containing information that could often
be qualified by diverse, numerous and complex definitions or terms, and
would reflect different levels of information.

Secondly, the system should place no restraints on the evolution of
the internal data base structure. This criterion implies that the system
should be capable of implementing new data files rapidly, and that fast
format re-organisation of already populated data files could be carried
out without the re-digitisation of information already stored under the
old format.

Thirdly, the system should be capable of interfacing both efficiently
and easily with application computer programs written by the Service.
This further implies that any internal re-structuring of portions of
the data base should not necessitate the revision of application programs
that access data not affected by the restructuring.

Fourthly, the system should be interactive so that the Service members and
authorised outside users may retrieve data by direct enquiry. The interactive portion of the system should be easy to use by the non-technical
(in the computer sense) enquirer and be capable of producing user formatted output.

Finally, the system must be easily accessible to all potential users
and stable - i.e. it has a guaranteed availability in the long term
and is fully supported by its designers.

The Choice

Data Banking is being carried out using the United States
Geological Survey (USGS) interactive computer system (Honeywell-Multics)
at Reston, Virginia, where three data base management systems are available
for use, namely; PACER, MIDS and MRDS. Of these PACER (Cargill, Olson,
Medlin and Carter, 1976) was written by the USGS to organise its National
Coal Resource Data System (Carter, 1976) (NCRDS) whereas MIDS (Multics
Integrated Data Store) and MRDS (Multics Relational Data Store) were
developed by Honeywell to run on their Multics operating systems (Honeywell, 1977). Of the three available choices the Service has concluded that
the data base management system best meeting its requirements was MRDS.

The main reason for choosing MRDS was that the relational data base approach
provides for a very high degree of data independence (Martin, 1977). Consequently, selected restructuring of the data base does not affect the
remaining data base structure and ensures that application programs are
relatively insensitive to those changes. This meets two of the Service's
requirements; that is, the data base can evolve without forcing massive
revision of application programs or unnecessary re-digitisation of data
already banked.

In addition the relational approach allows the data base designer to break
down or reduce complex data relationships into a series of simple tabular
relations containing directly associated data (Martin, 1977). Such simple
relations can be interrogated directly and independently of each other or
be linked together to carry out more complex retrievals. Retrievals,
additions and modifications of data are carried out using a Honeywell
supplied user 'front end' called LINUS (Logical INquiry and Update System).
This user interface provides a powerful high level non-procedural language
capable of being understood and used by individuals who are not experienced
in using computer systems. It also possesses a report generator facility
which may be used to output information formatted to user specifications.

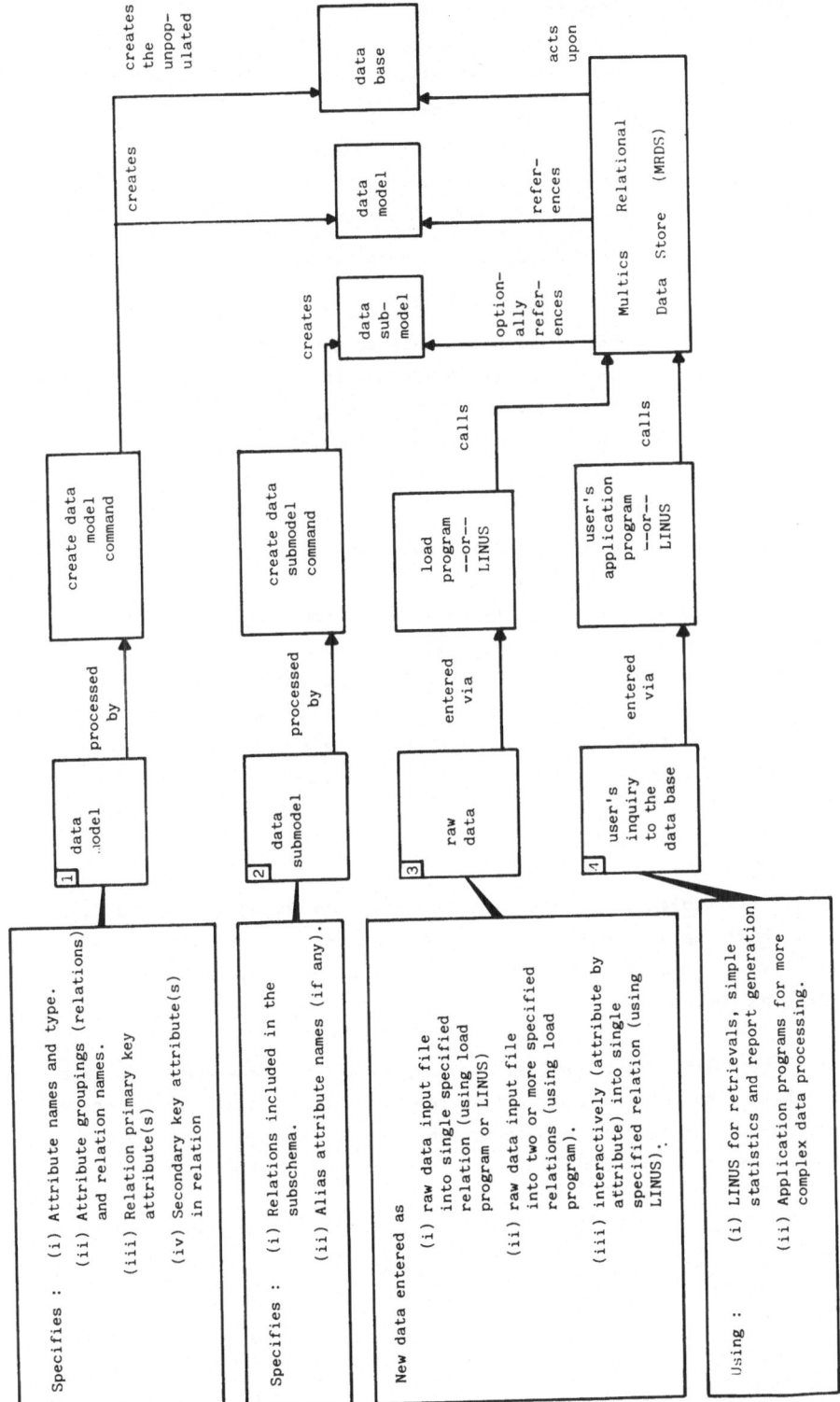

Specifies : (i) Attribute names and type.

(ii) Attribute groupings (relations) and relation names.

(iii) Relation primary key attribute(s)

(iv) Secondary key attribute(s) in relation

Specifies : (i) Relations included in the subschema.

(ii) Alias attribute names (if any).

New data entered as

(i) raw data input file into single specified relation (using load program or LINUS)

(ii) raw data input file into two or more specified relations (using load program).

(iii) interactively (attribute by attribute) into single specified relation (using LINUS).

Using : (i) LINUS for retrievals, simple statistics and report generation

(ii) Application programs for more complex data processing.

Fig. 4. Schematic breakdown of data base operations.

153

The prodcedure of creating and accessing a MRDS data base may be summarised as follows (see also figure 4):

(i)   Create a MRDS data model (or schema) which defines the characteristics and organisation of all the data within the data base.

(ii)   Create the unpopulated data base.

(iii)   Load unpopulated data base.

(iv)   Create (an) optional data submodel (s) or sub-schema (e) which define alternative (and often incomplete) descriptions of the existing data base. (Sub-schema may be optionally provided at any time by users to enhance "data independence").

(v)   Access the populated data base.

The MRDS data base operated by the Service is currently made up of eighteen individual relationships; these are shown schematically in figure 5. It is worth noting at this point that provision has been made for a set of lexicon 'files' or tables. These are composed of three columns; namely, term, country where used and term definition. Hence whenever a new term is introduced into a data base relation it can also be appended to the lexicon files together with its definition. This process serves two functions.

Firstly, it enables a comprehensive lexicon to be compiles as the data base is built up and secondly, it allows term definitions to be automatically appended as a footnote to listings of retrieved information containing those terms.

It is most unlikely that any user enquiry would require a search to pass through all the relations present in the data base. Usually the user would conduct an enquiry session by specifying a subschema made up of a subset of the relations containing pertinent information.

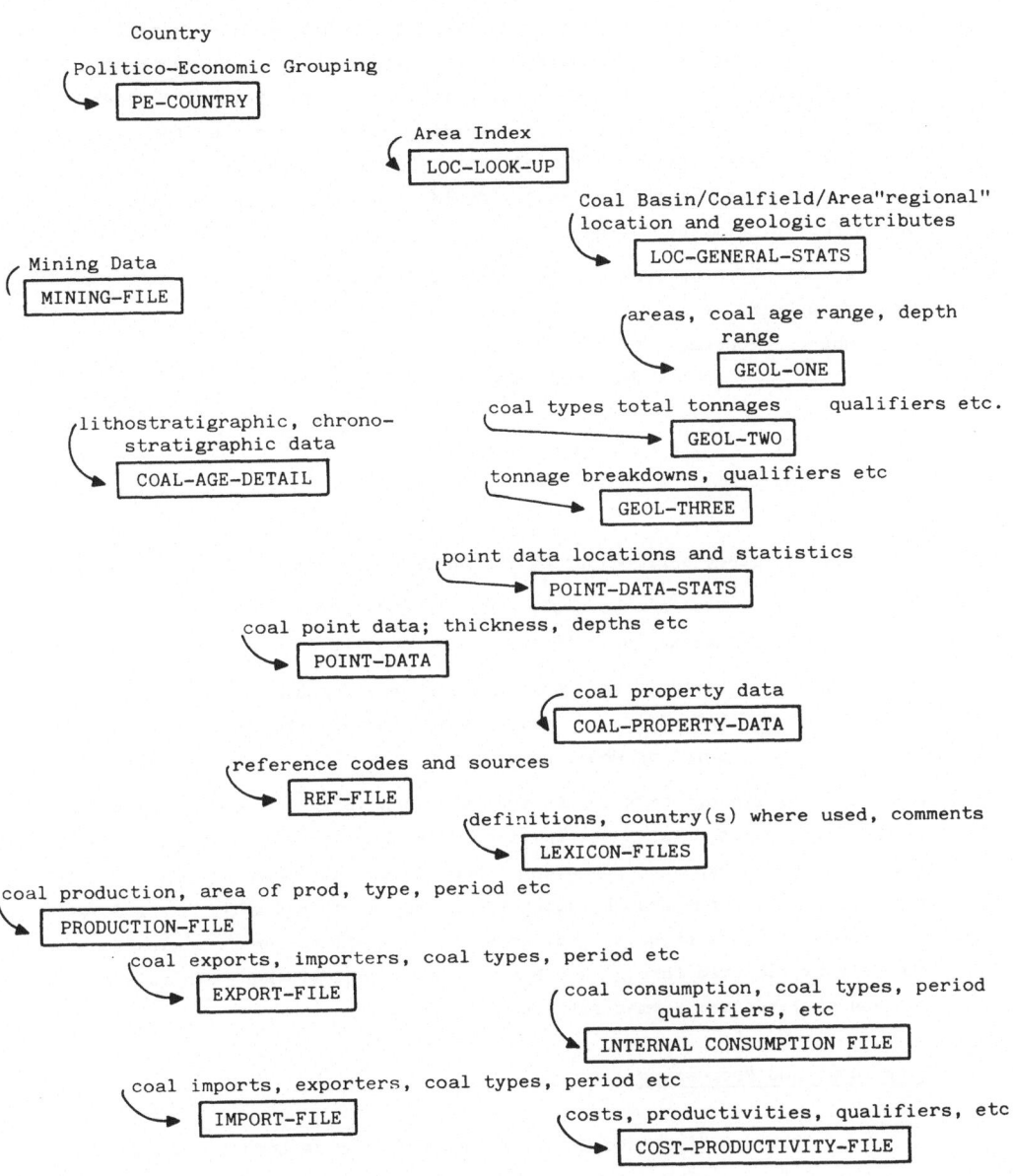

Country
Politico-Economic Grouping
PE-COUNTRY

Area Index
LOC-LOOK-UP

Coal Basin/Coalfield/Area"regional"
location and geologic attributes
LOC-GENERAL-STATS

Mining Data
MINING-FILE

areas, coal age range, depth
range
GEOL-ONE

lithostratigraphic, chrono-
stratigraphic data
COAL-AGE-DETAIL

coal types total tonnages   qualifiers etc.
GEOL-TWO

tonnage breakdowns, qualifiers etc
GEOL-THREE

point data locations and statistics
POINT-DATA-STATS

coal point data; thickness, depths etc
POINT-DATA

coal property data
COAL-PROPERTY-DATA

reference codes and sources
REF-FILE

definitions, country(s) where used, comments
LEXICON-FILES

coal production, area of prod, type, period etc
PRODUCTION-FILE

coal exports, importers, coal types, period etc
EXPORT-FILE

coal consumption, coal types, period
qualifiers, etc
INTERNAL CONSUMPTION FILE

coal imports, exporters, coal types, period etc
IMPORT-FILE

costs, productivities, qualifiers, etc
COST-PRODUCTIVITY-FILE

Fig. 5.   Schematic Representation of Data Base Relations

A typical subschema specification might be that as shown in figure 6.
Here only two relations are linked. The first (LOC-LOOK-UP) acts as
an area index and allows information given at various levels of areal
detail to be banked together with its areal associations. Using these
two relations the user can, say, instigate a search for borehole information within a specified area. For example, the user can specify that
he wishes to have a printout for all borehole names, year drilled and
geodetic locations in the Kempen Coalfield. This may be accomplished
in the following manner using LINUS.

```
select   locname year-drilled latitude longitude
from POINT-DATA-STATS
where    locname = (select locname
                    from LOC-LOOK-UP
                    where coalfield = "Kempen" &
                    loc-id = "borehole")
process
print
```

The search strategy carried out by the above expression is (see figure 6)

(a)   Find all borehole names in LOC-LOOK-UP which
            are assigned to the Kempen Coalfield.

(b)   Having found those names search through POINT-
            DATA-STATS and select specified values for
            printing where names match.

The method of linkage between the two relations in this example is defined
by the form of retrieval specification. The simple nested retrieval shown
can of course be expanded for other subschema to take into account much
more complex link and search patterns, and may be stored after use and
re-invoked during subsequent user sessions. In all instances a linkage
can only be achieved through a LINUS retrieval request when both relations
contain at least one common attribute.

Data Input and Processing

Data is digitised and validated off-line in London and stored
on cassette tapes. The digitised data is then transmitted via a 30

156

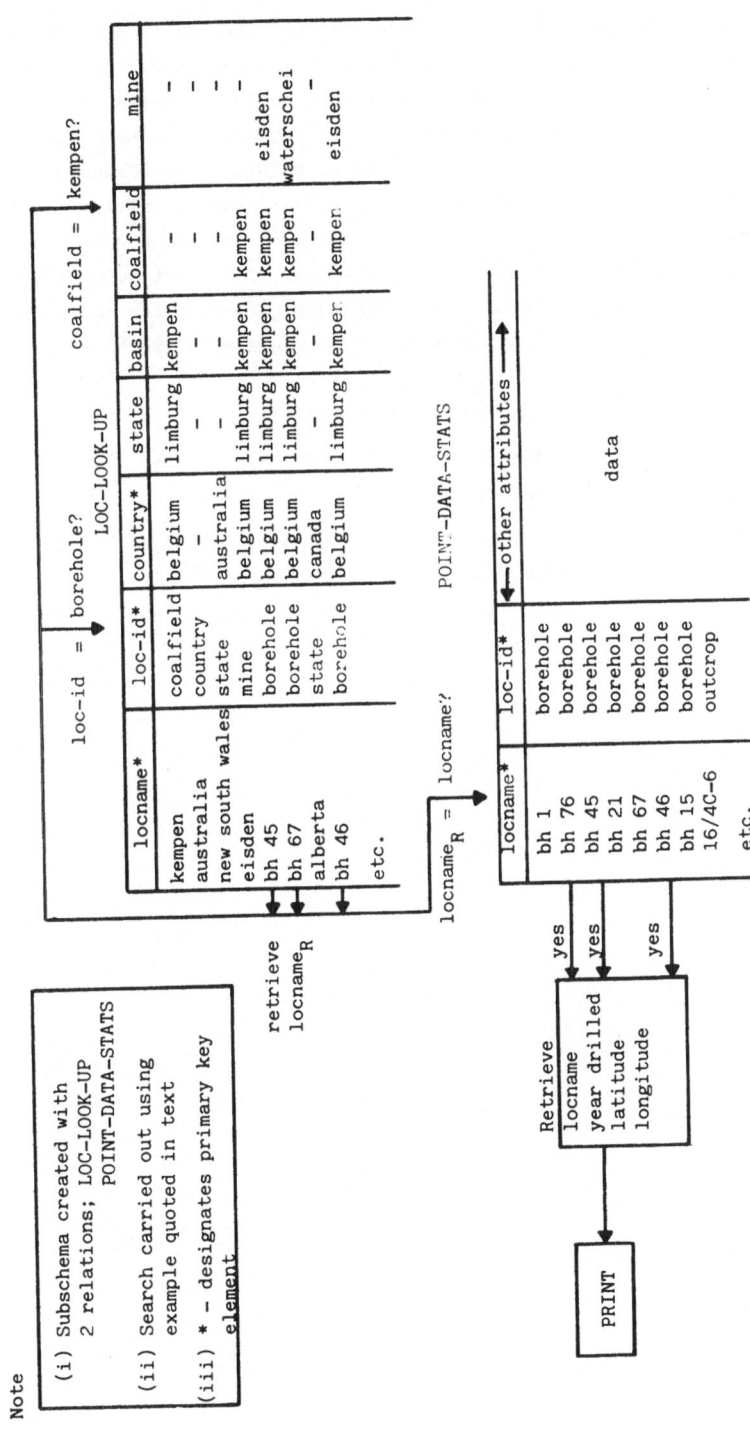

Note

(i) Subschema created with 2 relations; LOC-LOOK-UP POINT-DATA-STATS

(ii) Search carried out using example quoted in text

(iii) * - designates primary key element

Fig. 6. Search Strategy for a simple subschema

157

character/sec transatlantic link into computer storage as a raw data
input file (a 120 character/sec facility willbecome operational this
year). Where a raw data file contains entries for a single relation
only they may be entered directly using a LINUS option. More complex
input data files containing information to be stored simultaneously in
two or more relations is entered via a loading program written for that
purpose. Such programs, which use system supplied routines to access
the data base, are short and quickly written. This last point is
important as it allows input forms to be designed to match abstracted
information content without causing undue programming overheads.

Application programs can access data in two ways:

(a)   Via a data file created by LINUS from a
          retrieval request.

(b)   Directly, by embedding system supplied routines
          into the application program source code.

References

Cargill, S.M., Olson, A.C., Medlin, A.L., Carter, M.D. 1976.
PACER – Data Entry, Retrieval, and Update for the National Coal
Resources Data System (Phase 1).   Geological Survey Professional
Paper 978 : 107pp.

Carter, M.D. 1976. The National Coal Resources Data System of
the U.S. Geological Survey. Computers and Geosciences 2 : 337-340.

Clough, C.R. 1974. ASSASSIN, A Computer Based Information Storage
and Retrieval System. Paper presented to the Special Libraries
Associated Conference, Toronto, Canada.

Clough, C.R., Kilvington, L.C. 1978. ASSASSIN : the quiet revolution.
Program 12 : 35-47.

Honeywell Information Systems. 1977. Multics Relational Data Store
(MRDS) – Reference Manual. &c. No. AW53, Rev. 0.

IEA Coal Research 1977.   International Coal Projects.
London, &c. 20pp.

IEA Coal Research 1978.   IEA 1977 Coal Research Report.
London, &c. 16pp.

Martin, J. Author. 1977. Computer Data Base Organisation.
2nd ed. New Jersey : Prentice-Hall, 713pp.

OECD. 1977. Energy-Research, Development and Demonstration
Programme of the IEA. Paris, &c. 24pp.

This paper is based on work carried out by the Technical Information
Service and the World Coal Resources Data Bank Service. The author
gratefully acknowledges the assistance he has received from interested
staff members, particularly Ms Pam Harter of TIS. It should be understood, however, that the view expressed are those of the author.

INTERNATIONAL ACCESS TO THE PETROLEUM ABSTRACTS

INFORMATION SYSTEM

ROY W. GRAVES and JOHN A. BAILEY

Information Services Division, The University of Tulsa

1133 North Lewis Avenue, Tulsa, Oklahoma 74110

Summary: The Petroleum Abstracts Information System, which has operated since 1961, consists of two services: the weekly current awareness bulletin, Petroleum Abstracts, and the information retrieval service of printed indexes and magnetic tapes. Both services are supported by subscriptions from 53 major subscribers in the United States and in 18 other countries, and by 148 minor subscribers, mostly in the United States. In addition to this widespread coverage, some major subscribers also supply materials of the Petroleum Abstracts System to their offices in various localities around the world. Since October 1975 access, on an international basis, has been through the System Development Corporation (SDC) computer in Santa Monica, California. Major European points of access are London and Paris. Some major subscribers outside the United States use their own or Tulsa University computer programs locally to search the Master Record Tapes. Currently the level of search time for the SDC TULSA File is approximately 90 hours of computer time per month, of which 21 percent originates outside the United States. When the EURONET and SCANNET Systems become operative, an evaluation will be made of possible impact upon the present structure of the retrieval capabilities of the Petroleum Abstracts System. In the future, a more complete coverage of the literature of energy resources (wind, ocean currents and thermal differences, solid wastes, etc.) technology and economic minerals will be offered by the Petroleum Abstracts Information System.

INTRODUCTION

The Petroleum Abstracts Information System is designed and operated to aid the exploration and production activities of the petroleum industry. It functions from a small (13 full-time and approximately 17 part-time staff) non-profit organization, the Information Services Division, within the College of Engineering and Physical Sciences at The University of Tulsa. The annual budget of approximately $600,000 is derived predominantly from the fees paid by the 53 major subscribers to the services. In 1977, for the first time in the 17 year history of the System, approximately 50 percent of the operating capital came from sources outside the United States.

Two services are supplied by the System: the weekly current awareness bulletin, Petroleum Abstracts; and information retrieval services

consisting of indexes and magnetic tapes. Major Subscriber costs for each
service are shown in the following tabulation:

PETROLEUM PRODUCING COMPANIES

		Per Year
Group	I...Assets in excess of $1.5 billion......	$10,000
Group	II...Assets between $1.0 and $1.5 bil......	9,000
Group	III...Assets between $400 mil.& $1.0 bil....	8,000
Group	IV...Assets between $200 mil.& $400 mil....	7,000
Group	V...Assets between $100 mil.& $200 mil....	4,000
Group	VI...Assets between $ 50 mil.& $100 mil....	2,000
Group	VII...Assets between $ 10 mil.& $ 50 mil....	1,000

PETROLEUM RELATED COMPANIES

Group	I...Assets in excess of $1.5 billion.....	$ 4,000
Group	II...Assets between $1.bil. & $1.5 bil....	3,000
Group	III...Assets between $500 mil.& $1.bil.....	2,000
Group	IV...Assets between $ 10 mil.& $500 mil...	1,000

The Subscription rates shown in the above listing have not changed for
Petroleum Abstracts since inception of this service at the beginning of
1961. Costs for the information retrieval services have not changed since
the beginning of the present computer-based program in 1965. This is
partly attributable to cost saving modifications in the System but mainly
due to a continuing increase in the number of major subscribers.

In addition to the 53 major subscribers (petroleum companies and
petroleum service companies) the System is supported by subscriptions (to
the amount of about $37,000) from 148 minor subscribers (government agencies,
institutions, universities, and individuals). Approximately 650 serial publications provide the base of operations for the system. Articles from
these journals, state and national periodicals, published symposia, patent
disclosures or gazettes, etc., are selected for inclusion in the System on
the basis of pertinence to the petroleum exploration, production, and
development activities of the petroleum industry.

Selections of articles and patents for inclusion in Petroleum
Abstracts and the indexing of information contained therein are undertaken
by experienced scientists and engineers. Although most of the work is performed in-house some selected non-English language articles are dealt with
on a contract basis.

THE SYSTEM

Activities of the Information Services Division are concerned with
the daily production and maintenance of the two components of the System,
Petroleum Abstracts and the information retrieval materials and services.

Petroleum Abstracts

Each page-plate for the weekly bulletin is prepared "in-house" for
direct offset reproduction by a contract printer who also produces the
covers and mails subscription copies. Page format for the bulletin has
been changed from its original four abstracts per page to the present
column format (Figure 1A). Abstracts average about 150 words. Particular
attention is paid to the presentation of maximum information in a readable
style. Two hundred to 300 abstracts are issued each week, and by the end
of 1977 more than 240,000 abstracts were published. Back issues of
Petroleum Abstracts are available on microfilm

Abstracts are presented in classified order:
    GEOLOGY
    GEOCHEMISTRY
    GEOPHYSICS
    DRILLING
    WELL LOGGING
    WELL COMPLETION & SERVICING
    PRODUCTION OF OIL AND GAS
    RESERVOIR ENGINEERING & RECOVERY METHODS
    PIPELINING, SHIPPING & STORAGE
    ECOLOGY & POLLUTION
    ALTERNATE FUELS & ENERGY SOURCES
    SUPPLEMENTAL TECHNOLOGY
    MINERAL COMMODITIES

Information Retrieval Materials and Capabilities

Information contained in the articles for which abstracts are
published in Petroleum Abstracts is indexed manually from the Exploration
and Production (E & P) Thesaurus, the Geographic Thesaurus, and from the

GRAVITY INTERPRETATION   246,084

FOURIER TRANSFORMS OF FINITE LENGTH THEORETICAL GRAVITY
ANOMALIES-- R.D.Regan (US Geological Survey) and W.J.Hinze
(Purdue Univ); GEOPHYSICS v. 42, No. 7, pp 1450-1457,
Dec. 1977
        The mathematical structure of the Fourier
transformations of theoretical gravity anomalies of several
geometrically simple bodies appears to have distinct
advantages in the interpretation of these anomalies.
However, the practical application of this technique is
dependent upon the transformation of an observed gravity
anomaly of finite length. Ideally, interpretation methods
similar to those for the transformations of the theoretical
gravity anomalies should be developed for anomalies of a
finite length. However, the mathematical complexity of
the convolution integrals in the transform calculations of
theoretical anomaly segments indicate that no general closed
analytical solution useful for interpretation is available.
Thus, in order to utilize the Fourier transform
interpretation method, the data must be of sufficient
length for the finite transform to closely approximate the
theoretical transform. (19 refs.)

A. Typical Abstract

CALCAREOUS ALGA
- UF   CORALLINACEAE
- UF   CORALLINE ALGA
- BT   ALGA
-         THALLOPHYTA
- SA   BOTANY
- SA   MEGAFOSSIL
- SA   MEGAORGANISM
- SA   MICROFOSSIL
- SA   MICROORGANISM
- SA   PALEONTOLOGY
- SA   PLANT (BOTANY)
- SA   REEF BUILDER

CALCAREOUS DEPOSIT
- **   (65-66) INDEX ALSO AS
-         SEDIMENT(S) (GEOLOGY),
-         SEDIMENTARY ROCK(S),
-         AS APPLICABLE.
-         NT (65-69) CALCAREOUS MUD
-         NT (67-69) CALICHE
-         NT (67-69) LIME MUD (GEOLOGY)
-         BT (67--) DEPOSIT (GEOLOGY)
- UF   CALCAREOUS MUD
- UF   CALCAREOUS ROCK
- UF   CALCAREOUS SEDIMENT
- UF   CALCAREOUS STRATA
- BT   DEPOSIT (GEOLOGY)
- SA   CALICHE
- SA   LIME MUD (GEOLOGY)
- SA   SEDIMENT (GEOLOGY)
- SA   SEDIMENTARY ROCK
- SA   TRAVERTINE

CALCAREOUS MUD
- **   USED (65-69).
-         BT (65-69) CALCAREOUS
-         DEPOSIT(S)
- USE   CALCAREOUS DEPOSIT

B. E & P Thesaurus

JEKYLL ISLAND
- **   ADDED APRIL 1965
- BT   GEORGIA
-         EASTERN US
-         UNITED STATES
-         NORTH AMERICA

JEMEZ MT
- **   ADDED NOVEMBER 1966
- BT   NEW MEXICO
-         WESTERN US
-         UNITED STATES
-         NORTH AMERICA

JEMMAPES AREA
- **   ADDED OCTOBER 1970
- BT   ALGERIA
-         AFRICA

JENNINGS RIVER MAP AREA
- **   ADDED OCTOBER 1969
- BT   BRITISH COLUMBIA
-         CANADA
-         NORTH AMERICA

JEQUITINHONHA DELTA
- **   ADDED JUNE 1972
- BT   BRAZIL
-         SOUTH AMERICA

JERICO AREA
- **   ADDED APRIL 1976
- BT   ISRAEL
-         ASIA

C. Geographic Thesaurus

Figure 1.

lists of Supplementary Descriptors. This process produces from 10 to
more than 100 descriptors for each article. The manual indexing furnishes
the basic input for production of the information retrieval materials and
services. A computer program provides autoposting of descriptors from
the hierarchies of both thesauri. Thus, only the most specific term for
any concept is posted manually. All broader terms are added automatically.

Exploration and Production Thesaurus

The Exploration and Production (E & P) Thesaurus was prepared as a
cooperative effort between the Information Services Division and the major
subscribers. The current edition, the fifth, is the product of extensive
revision and updating. An excerpt from the E & P Thesaurus is shown in
Figure 1B.

Geographic Thesaurus

Terms (descriptors) included in the Geographic Thesaurus are limited
to sedimentary basins, geographic features, and selected geographic areas.
The geographic descriptors listed are those assigned by the indexers; ultimate authority for location is the original article from which the term was
derived. A revised and expanded third edition of the Geographic Thesaurus
is available. The Geographic Thesaurus is patterned after the E & P Thesaurus. Figure 1C illustrates the entries.

Alphabetic Subject Index

The Alphabetic Subject Index (ASI) is a manual search index designed
for the rapid location of major articles and patents on a given subject,
for review(s) of current technology, and for browsing. This Index brings
together, under a single subject heading, titles of articles and patents
whose abstracts have appeared in the weekly issue of Petroleum Abstracts.
The ASI is photocomposed from processing tapes and is printed and distributed bi-monthly, with the final issue being a hard-bound, 12 months' cumu
lative index. Three appendices contain bibliographic information (arranged numerically by abstract number), an alphabetic listing of authors, and
a patent index (arranged sequentially by patent or application number,
under each individual country). Typical entries in the Alphabetic Subject Index are shown in Figure 2A

A. Alphabetic Subject Index

OIL AND GAS ORIGIN
(181)

```
•140,590 •138,461 •139,442 •140,983 •138,694 •140,325 •138,486 •139,917 •
•140,650 •138,471 •140,572 •143,313 •138,944 •140,615 •138,696 •140,597 •
•142,730 •140,571 •142,972 •143,453 •140,264 •140,625 •140,276 •142,737 •
•144,070 •143,071 •143,032 •143,843 •141,564 •143,845 •142,736 •143,367 •
•145,090 •144,841 •143,362 •144,473 •143,354 •144,065 •143,356 •145,797 •
•145,130 •145,131 •144,802 •144,883 •143,364 •144,745 •145,826 •150,157 •
•145,810 •145,811 •147,572 •145,803 •145,824 •146,735 •146,096 •151,317 •
•147,560 •146,751 •149,922 •147,803 •147,794 •147,575 •146,746 •151,697 •
•147,800 •146,761 •150,392 •149,143 •148,364 •147,775 •146,766 •152,747 •
•149,720 •149,111 •151,692 •149,453 •149,464 •148,085 •147,546 •        •
•150,180 •149,931 •151,982 •149,763 •150,134 •148,395 •147,566 •        •
•150,550 •150,181 •152,412 •149,943 •150,184 •148,655 •148,696 •        c
•152,570 •150,391 •152,442 •150,183 •151,324 •149,145 •149,456 •        •
•153,720 •151,341 •154,052 •151,103 •151,984 •149,465 •149,716 •        •
•154,310 •152,431 •154,282 •152,433 •152,374 •150,405 •150,396 •        •
•        •153,061 •154,292 •154,273 •154,264 •151,115 •150,406 •        •
•        •153,071 •154,312 •154,333 •        •151,325 •151,126 •        •
•        •153,191 •154,332 •        •        •151,695 •151,696 •        •
•        •153,351 •154,862 •        •        •153,055 •152,946 •        •
•        •154,291 •154,912 •        •        •154,005 •153,056 •        •
•        •154,311 •        •        •        •154,045 •153,086 •        •
•        •154,361 •        •        •        •154,245 •154,055 •        •
•        •        •        •        •        •154,285 •154,256 •        •
•        •        •        •        •        •154,305 •154,356 •        •
•        •        •        •        •        •154,315 •154,906 •        •
•        •        •        •        •        •154,515 •        •        •
```

B. Dual Dictionary Coordinate Index

Figure 2.

Dual Dictionary Coordinate Index

The original concept of the Dual Dictionary (DD) Coordinate Index was for the issuance each year of two printed copies at 4-month intervals, cumulative at 8 and 12 months. The Dual Dictionary is distributed on 16-mm microfilm in July and December, the latter issue being cumulative for the year. The Index also is available on microfiche.

Two copies of the DD constitute the Dual Dictionary concept. Each copy contains an alphabetical listing of all descriptors used in indexing the information contained in the articles for which abstracts are published in Petroleum Abstracts. The abstract numbers are arranged by terminal digit in order to facilitate coordination. Each abstract number is listed under every descriptor that was assigned by the indexer or that was generated from the E & P Thesaurus and Geographic Thesaurus hierarchies, by computer processing. With the present microfilm or microfiche format, copies of the pertinent DD page entries may be made with a reader-printer, to facilitate coordination. An example from a printed copy of the DD is shown in Figure 2B.

Master Record Tapes

When complex searches are needed, computer searching becomes feasible and economic. For this purpose, computer search tapes, called Master Record Tapes, are issued every four months. These tapes are organized sequentially by abstract number and contain all the descriptors assigned or generated (by autoposting from E & P Thesaurus and Geographic Thesaurus hierarchies) for each article concerned. The tape can be searched sequentially with available computer programs. It is issued in several formats to match the requirements of the subscriber's specific computer. Some search systems are available on local or national/international retrieval systems.

Additional Materials and Services

Certain other materials and services are available in addition to those already described:

Supplementary Descriptors Lists - some specific terms such as chemical names, company names, geographic and geologic named features
(excluded from the Geographic Thesaurus), and E & P Thesaurus-
type descriptors (added to the System as needed) are contained in
the lists of Supplementary Descriptors.

Key Word out of Context (KWOC) List - this is a list of descriptors
from the E & P Thesaurus and the added E & P Thesaurus-type descriptors that are organized in KWOC format so that those related
to a particular subject (or concept) are listed together.

Related Patent Index - this is a list of each related patent processed since January 1, 1973 (issued at six month intervals and
cumulative from January 1, 1973). The index is ordered by the
primary patent abstract number to which the patent is related
and contains, as well, the related patent abstract number with
its country of issue and patent number.

Card Files -

1. Classification Retrieval System - During the period 1961-1964
   the System used classification coding for information retrieval. This file can be searched manually on request. It
   contains, in addition to the System input, other classification card files that go back for more than 30 years (donated
   to the System by a major oil company).

2. Author File - For purposes of retrieving information based
   on author names and for checking against previously published articles or abstracts, a card file of authors is maintained. Abstract cards are used for this purpose.

3. Accession Number File - Since January, 1965 (beginning with
   abstract number 50,001) an abstract card accession number
   file has been maintained. This file provides quick access to
   any numbered abstract and facilitates easy reproduction of
   abstracts.

4. Patent File - A patent file (of abstract cards) is maintained for retrieval purposes and for checking for related patents.
   This file is separated by country and contains both patent
   number and priority date listings.

Computer Hardware and Processing

The computer used by The Information Services Division is a Xerox

Sigma 6, located at the University's Computer Center. All computer programs for processing and handling in the System are written in Fortran IV with a number of important assembly language subroutines. Thirty-plus programs are needed for computer processing in the System.

INTERNATIONAL ASPECTS OF THE SYSTEM

Worldwide Distribution of Materials

Petroleum Abstracts and information retrieval materials (indexes, tapes, etc.) are used in 47 countries around the world. Nine countries (including the US) in North, South, and Central America have access to the System, as do 13 in Europe, 6 in the Middle East, 9 in Africa, and 10 in the Far East and Australia. This distribution is presented in Table 1, which shows also a listing of major oil company offices outside the United States that receive the System materials. The listing here is for those offices to which materials are sent from Tulsa. Some companies forward materials to the same or similar locations, from their own headquarters. Tulsa mailings are to 29 countries outside the US, and in 17 of these, this is the only source in that country for Petroleum Abstracts Information System materials.

Offices of major and minor subscribers in all of the 47 countries receive Petroleum Abstracts. Eight major subscribers in the US, 4 in Canada, 2 in Europe, 3 in Asia and 1 each in Africa and South America do not receive the information retrieval indexes and magnetic tapes. Sixty of the 148 minor subscribers do not receive information retrieval indexes. Information searches for subscribers who do not receive the retrieval materials can be made on request from the System's in-house capabilities (both computer and manual). In many instances search requests from the US and other countries are received in Tulsa by mail, teletype, and telephone. Such searches are performed on a standard fee basis, the results being mailed to the requestor.

Petroleum Abstracts Search System

The Petroleum Abstracts Search System (PASS) is a coordinated set of computer routines that provides the user with a comprehensive literature search capability. Operating on-line from a remote terminal, a user

Distribution of PETROLEUM ABSTRACTS and Related
Services

| Country | Number Subscribers | | Major Company Offices * |
	Major	Minor	
AMERICAS	36	111	16
Canada	7	22	11
United States	24	89	-
Mexico	1	-	-
Trinidad	-	-	1
Venezuela	1	-	1
Colombia	1	-	-
Brazil	1	-	-
Peru	1	-	-
Argentina	-	-	3
EUROPE	11	25	20
Sweden	-	1	-
Finland	-	1	-
Norway	1	1	2
United Kingdom	3	9	10
Ireland	-	-	1
Netherlands	-	1	-
Belgium	1	2	2
West Germany	2	7	1
France	3	-	2
Switzerland	-	2	-
Austria	-	1	1
Italy	1	-	-
Spain	-	-	1
MIDDLE EAST	3	0	4
Turkey	1	-	1
Iran	-	-	1
Kuwait	1	-	-
Saudi Arabia	1	-	-
Qatar	-	-	1
Muscat and Oman	-	-	1

Table 1

Distribution of PETROLEUM ABSTRACTS and Related
Services

Country	Number Subscribers		Major Company Offices *
	Major	Minor	
ARFICA	1	4	9
Tunesia	1	-	1
Libya	-	-	3
Egypt	-	-	2
Ivory Coast	-	1	-
Nigeria	-	1	1
Gabon	-	-	1
Congo	-	-	1
Tanzania	-	1	-
South Africa	-	1	-
FAR EAST and AUSTRALIA	2	8	22
Pakistan	-	-	2
India	1	-	-
South Korea	-	1	-
Japan	1	-	1
Philippines	-	1	-
Brunei	-	-	1
Malaysia	-	-	1
Singapore	-	-	7
Indonesia	-	-	5
Australia	-	6	5
TOTAL (47 countries)	53	148	71 (29 Countries outside US-only distr.in 17 countries)

* Materials mailed from Tulsa

Table 1 (Cont'd)

can structure a search, review the results, and restructure as necessary to obtain whatever amount of generality or specificity is desired. Functions are available to allow display of results on the terminal and to print the final search results on the Computer Center line printer.

The search program is supported by two major information bases: an inverted search file of descriptors that lists all abstract numbers to which each descriptor has been assigned; and the item files containing titles, bibliographic citations, and author names. Physical files in each class are stored on separate disc-packs for literature and for patents and are updated monthly.

ORBIT IV Search System

Through an arrangement with System Development Corporation (SDC), the ORBIT IV Search System adds a new dimension to the Petroleum Abstracts Information System. The information base of the Petroleum Abstracts System (literature and patents) now can be accessed through SDC's Search Service ORBIT IV Search System (file TULSA).

The ORBIT IV Search System is an extensive and easily used information retrieval program that has been in operation on a daily basis since 1970. The program allows the searcher to specify his information needs by a logical combination of descriptors and substantive keywords drawn from the titles of papers. Searches also can be made on specified author names. An interactive system, ORBIT IV provides capability for progressively refining the statement of a search request to quickly and easily identify the item of primary interest. The TULSA file is updated quarterly.

Search results can be printed on the user's terminal or they can be directed to a high speed, off-line printer. Off-line reports are delivered by airmail from System Development Corporation's offices in Santa Monica, California.

ORBIT IV is accessed through a worldwide communications network (Tymshare) that provides local telephone access to minicomputers in major US population centers as well as selected cities in Canada and Europe. The System also may be used directly through WATS lines or by direct distance dialing. Venezuela uses a slow speed cable to access a Tymshare point in New York; Mexico uses long distance telephone to a Tymshare point in Texas.

The ORBIT IV Search System provides access to many other information files, some selected examples of which are as follows:

American Petroleum Institute (APILIT/APIPAT)

American Chemical Society (CHEMCON)

American Geological Institute (GEOREF)

Smithsonian Scientific Information Exchange (SSIE)

National Technical Information Service NTIS)

The cost of searching the Petroleum Abstracts information base is based on an hourly charge for use of the file, an hourly charge for the communications network (optional), and a charge per reference for off-line printing. The actual rate varies with the status of the subscriber. The present level of search time for the TULSA file is approximately 90 hours of computer time per month. About 21 percent of this searching orginates outside the United States.

The TULSA file on ORBIT IV, can be accessed only by the user with a search license. Licenses are issued automatically with subscriptions to the System's information retrieval services. In addition, some information brokers have purchased search licenses. Two major subscribers to the services have purchased a search license only. The fee for this category of service is approximately 15% less than the fee for a complete subscription to the retrieval services.

European Information and Data Networks

Preliminary talks concerning possible and/or feasible access to the EURONET and SCANNET systems have been held with several European subscribers. Any definitive action will have to await completion of the networks.

European subscribers to the Petroleum Abstracts Information System access the SDC TULSA file predominantly through Tymshare stations in London and Paris. Most European usage, to the present time, of the SDC TULSA file initiates in the United Kingdom, the Netherlands, France, and Norway.

Some American information systems are reported to have commitments to the EURONET system (Ungerer, 1977 :205). Several of them are part of the SDC ORBIT IV Search System. Inasmuch as an apparent compatability exists between the SDC and EURONET information base systems, no technical deterrent would obstruct entry of the Petroleum Abstracts Information System into the EURONET system. The same situation would apply to the proposed SCANNET system for which a close cooperation with the EURONET system is contemplated (Gronlund and Nilsson, 1977 :68).

At present the Petroleum Abstracts Information System is under an exclusive short term license agreement with the SDC Search Service. Any future arrangements for entry of the Petroleum Abstracts Information System retrieval files into any European system would have to be mutually beneficial to the System and to the major subscribers, and financing would be a major consideration.

SYSTEM EXPANSION

Because of the nature of the literature and patent base for the Petroleum Abstracts Information System, capability exists for rapid expansion into other energy and/or mineral fields. The present category of Alternate Fuels and Energy Sources covers the engineering, drilling, and development aspects (but not production) of in-situ coal gasification, tar sands, oil shale, and geothermal energy. Exploration aspects are covered in the exploration sections of the System (except for coal). Expansion into other energy sources such as wind, solar, tidal, waste materials, etc. is feasible and contemplated.

The Mineral Commodities section is concerned with exploration only and is confined to complete articles and patents of selected metals, nonmetals, and mineral-fuel deposits. Abstracts of the articles are not published although the title, author, and bibliographic citation is given. Indexing is shallow and limited to a maximum of 10 assigned descriptors.

The nonmetals and mineral-fuel deposits are coal, phosphate, potash, sulfur, and uranium-thorium. The metals mineral deposits include aluminum, chromium, cobalt, copper, gold, iron, lead, manganese, molybdenum, nickel, silver, tin, titanium, tungsten, vanadium, and zinc. This coverage can be expanded to whatever additional commodities are required as well as to mining, haulage, and basic ore beneficiation. At present, no expansion into milling, or other metallurgical practices is contemplated.

The possible expansion mentioned above could take place within the Petroleum Abstracts Information System format, but this does not preclude the eventual separation into several systems.

REFERENCES

GRONLUND, B. and NILSSON, N. 1977. Scannet, an I&D data network for the Nordic countries. In 1st International On-Line Information Meeting. Oxford: Cotswold Press : 65-69.

UNGERER, H. 1977. Euronet: A new comprehensive information utility for the European user. In 1st International On-Line Information Meeting. Oxford: Cotswold Press : 203-214.

BIBLIOGRAPHY ON PETROLEUM ABSTRACTS INFORMATION SYSTEM

ANONYMOUS 1967. "Petroleum Abstracts" provides key for speedy information retrieval. Oil and Gas Journal 65/17 : 104-105

BAILEY, J.A. 1972. An interactive literature search system for the XDS Sigma series computers. Presented at the 18th International XDS Users-Group Meeting : 11pp.

BRENNER, E H. and HELANDER, D.P. 1969. Petroleum literature and patent retrieval-centralized information processing. Special Libraries 60 : 146-152.

GRAVES, R.W., HELANDER, D.P. and MARTINEZ, S.J. 1969. The University of Tulsa information retrieval system. Geoscience Information Society Proceedings 1 : 18-26.

GRAVES, R.W. and HELANDER, D.P. 1970. A feasibility study of automatic indexing and information retrieval. IEEE Transactions on Engineering Writing and Speech EWS-13 : 60-64.

GRAVES, R.W. and BAILEY, J.A. 1977. Petroleum Abstracts information system. Geoscience Information Society Proceedings 7 : 72-94.

GUERRERO, E.T. and MARTINEZ, S.J. 1962. Preventing technical obsolescence in petroleum engineers and scientists. Journal of Petroleum Technology 14 : 931-934.

GUERRERO, E.T. and MARTINEZ, S.J. 1964. Industry blight: technological obsolescence. Petroleum Management 36 : 94-99, 161.

GUERRERO, E.T., MARTINEZ, S.J. and GRAVES, R.W. 1966. Methods of making petroleum engineers aware of new technology. Journal of Petroleum Technology 18 : 659-575.

GUERRERO, E.T., HELANDER, D.P. and MARTINEZ, S.J. 1968. What is being done by the petroleum industry to meet the information problem. Presented at the State Technical Services Institute Conference on the Petroleum Industry : 11 pp.

HELANDER, D.P., GRAVES, R.W. and MARTINEZ, S.J. 1967. Sources of "lost" information for log analysts. 8th Annual SPWLA Logging Symposium Transactions Paper G : 19 pp.

MARTINEZ, S.J., GRAVES, R.W. and HELANDER, D.P. 1967. Multi-level searching of the petroleum exploration and production literature. Proceedings of the American Documentation Institute 4 : 249-253.

MARTINEZ, S.J. and HELANDER, D.P. 1968. The development and maintenance of a specialized controlled vocabulary thesaurus. Proceedings of the American Society for Information Science 5 : 279-283.

MARTINEZ, S.J., ANIS, M., BUTHOD, P. AND HELANDER, D.P. 1969. Machine generation of information retrieval indexes. Proceedings of the 6th Annual National Colloquium on Information Retrieval 6 : 51-65.

MARTINEZ, S.J., BROWN, L.P., HELANDER, D.P. and MCLEOD, H.O. 1969. Computer processing of thesaurus data. Proceedings of the American Society for Information Science 6 : 269-275.

MARTINEZ, S.J. 1970. An information system for petroleum engineering technology. IEEE Transactions on Engineering Writing and Speech EWS-13 : 58-59.

MARTINEZ, S.J., and BAILEY, J.A. 1972. Revising magnetic tape-stored information retrieval files to attain descriptor compatability. Presented at the American Society for Information Science Meeting Preprint Booklet, Paper 7 : 5pp.

MARTINEZ, S.J. 1973. A cooperative information storage and retrieval system for the petroleum industry. Journal of Chemical Documentation 13 : 59-65.

MARTINEZ, S.J. 1973. The use of a permuted word list to facilitate indexing and searching. Proceedings of the American Society for Information Science 10 : 137-138.

MARTINEZ, S.J. and BAILEY, J.A. 1978. A historically documented thesaurus for improved retrospective information retrieval. Presented at the 175th American Chemical Society National Meeting. Anaheim, California : 14pp.

MCLEOD, H.O., GRAVES, R.W. and MARTINEZ, S.J. 1969. Economic minerals abstracts - the development of a technical information service for the mineral exploration & mining industry. Presented at the Fall Meeting, Society of Mining Engineers of AIME. Salt Lake City, Utah : 18pp.

WATTENBARGER, D.W., BAILEY, J.A. and MARTINEZ, S.J. 1977. Interactive system for controlled vocabulary maintenance. Presented at the <u>ACM '77 Conference</u>. Seattle, Washington : 7pp.

AN ADVANCED HYDROGEOLOGICAL DATA STORAGE AND RETRIEVAL SYSTEM

IN USE IN NEW SOUTH WALES

J.B. JOSEPH

Resources Division

Water Research Centre, Medmenham Laboratory, P.O. Box 16, Medmenham
Marlow, Bucks. SL7 2HD

R.J. GARRARD
D.R. WOOLLEY

Hydrogeologists with the Water Resources Commission,

Ibis House, Miller Street, North Sydney, New South Wales 2060, Australia

Summary: The Water Resources Commission of New South Wales holds records
of some 50 000 wells and boreholes. In 1968 the feasibility of an automated data storage and retrieval system for these records was studied;
this led directly to the design of the present computer based system which
comprises a core file of site and construction data and several peripheral
files relating to water quality, water levels, etc. Data handling programs
were written in Cobol, but Fortran can be used for manipulation if this is
more efficient. The data for several files are entered onto the coding
sheets at source, e.g. in the laboratory for quality data. All data are
checked before key-entry, and then proof read after processing. Data input
is via cassette tape, while output may be in one of a number of forms
including microfiche (computer output microform (COM)) and magnetic tape
formatted for use on other machines.

1. INTRODUCTION

New South Wales is the most populous state in Australia. It has an
area of 8×10^5 Km^2 and includes a broad range of physiographic and clima-
tic regions, from low relief and rainfall in the west to high relief and
rainfall on the eastern coast. The Water Resources Commission of New
South Wales is responsible for all aspects of groundwater control and
management in the State. A major part of that responsibility is 'the
maintenance of a central repository of relevant data on water resources
and on the use of those resources' (New South Wales Government, 1976).
Groundwater data are obtained by the Commission from two main sources:-

(i)   a licensing scheme which requires that all boreholes sunk for
       groundwater exploration or extraction must be licensed. Under
       the license conditions the Commission must be informed of strata
       penetrated, quantity and quality of water found, borehole
       construction details and site location. In the case of high-
       yielding boreholes (irrigation, industrial, or town water supply
       use) an annual statement of the amount of water extracted is
       required;

(ii)   the Commission's own regional groundwater investigations
       which include drilling, water level measuring, and water sampling.

177

The manual card system of data storage, maintained by the Hydrogeological Section of the Commission, was found to be unsuitable for the increasing amount of periodic data (water levels, chemical analyses, etc.) which was being collected from a rapidly growing network of observation boreholes in the mid 1960's. The system was also inadequate for statistical analyses of the data and for regional investigations, although it was well suited to small area investigations such as selection of a drilling site for an individual farmer.

In 1968 a feasibility study of the use of computer systems for the storage and retrieval of borehole data was undertaken by the Commission and the State Government Automatic Data Processing (ADP) Bureau. The results of this study were submitted as a Management Report (Lorimer, Woolley and Lowe,1970), incorporating a design for a computer based bore data handling system which formed the basis of the system described in the present paper. The overall design concept of a core (location/construction) file with a fixed format, and peripheral data files allowing for unlimited amounts of time dependent data, and the decision to use Cobol as the prime programming language for data storage and retrieval, stemmed from the ADP Bureau. The study also showed that any scheme adopted had to provide random manual access to the data, on the same basis as the access provided by the earlier card system.

The system is now in operation, although the core file is not yet complete. Random manual access is achieved by providing a combined listing, updated quarterly, of the location/construction and lithology files. This facility is necessary in order to deal effectively with requests for assistance from farmers, and is additional to statistical analysis of, and selective retrieval from, the stored data.

The system proposed in the feasibility study was adopted in 1970. In order to implement the scheme a data coding group comprising four people was set up, and a hydrogeologist was trained in programming and systems techniques. This group proceeded with the initial stages of implementation, with considerable programming assistance from the ADP Bureau. Since 1974 the group has been expanded to include in-house programming and key-entry staff. In its early life the system was proved and run on the State Government's Honeywell 800 computer, but it was converted to run under the 'Mod-8' operating system on the Honeywell 8200 when that became available in 1972. At present all data files are maintained on magnetic tape.

A major addition to the system became necessary in 1972, when the licensing requirements for high yielding boreholes were changed by the Commission.

The main changes were in the issuing of licenses for a fixed and limited term, and in the statutory requirement for an annual statement of

the volume of water pumped. The system devised to handle the licensing
data is separate from the main storage and retrieval system, but is
designed to enable data to be extracted from both systems concurrently if
required, and to allow the ready exchange of data between the systems.

2. THE SYSTEM

GENERAL PRINCIPLES

The general form of the system is illustrated in Fig. 1, and the
processing sequence for an individual item of data in Fig. 2. The method
of ordering the information on each tape file, and of identifying the
information referring to a particular borehole stems from the existing
borehole numbering and location-recording system. Boreholes are numbered
in a single continuous sequence, the order relying solely on the date of
notification of the data. Thus each borehole is defined uniquely by one
number of up to six digits. Borehole locations are recorded on a set of
maps, each map being sub-divided into a 16 square lattice. Borehole data
are held on the files in numerical order within each map area, the blocks
of data for each map area being ordered in map number sequence. In
addition to using the bore number and map number as a means of ordering
the data on the files, they are used together as a means of linking data
from the various files, the use of both numbers acting as a check on errors.

A 'grandfather/father/son' updating system is used for all files, as
shown diagramatically in Fig. 3. This allows the re-creation of the latest
master file in the unlikely event that it is corrupted or accidentally
destroyed. The great-grandfather is kept in remote storage as a further
safety precaution. To date, it has not been necessary to make use of these
back-up facilities.

Although the files cover specific subsets of information, they have a
number of features in common.

(i)   The data are stored entirely on magnetic tape, for historical
       reasons. The twin results of this are sequential storage, and
       the updating procedure referred to above.

(ii)   Each individual record contains identification fields in addition
       to the basic data. Basic identification data are the same
       for all files and comprise the borehole and map numbers as
       described above. Other identification information without
       which data cannot be entered includes for example: on the
       chemistry file the date of sampling, and on the core file
       the detailed location of the borehole.

(iii) In order to make programming easier all records are of fixed
       format. As this could lead to excessive areas of wasted
       space on the tapes, and to correspondingly lengthy

processing times, the data is reduced to a number of sub-records whose presence, or absence, is indicated by the use of a flag in the identification block of the record. Where a sub-record would be completely empty it is simply omitted from the record on the master file.

(iv) Information is stored entirely in metric units, although the system allows entry in either metric or imperial units.

The use of a common identification format for records on all files has resulted in two great advantages. Firstly, extraction programs can readily match data stored on any one file with that for the same borehole on any of the other files. For example, boreholes with low salinity water (chemistry file) in sandstone aquifers (lithology file) deeper than 300 metres (core file) could be listed. Secondly the system can be expanded quite easily in any desired direction; for example, a file devoted to pumping test data might be a useful addition when the more pressing work of creating the central data store is complete. In this context the use of Cobol as the language for the storage, retrieval and maintenance programs is of great assistance. Cobol excels in file-handling and linking, and should not be disparaged as 'non-scientific'; it is designed to cope with accountancy procedures, and the storage and retrieval of scientific data can be viewed as an accounting exercise. The place for Fortran, or any other scientific language, is in the analysis of assemblies of data after their retrieval and reorganisation.

CORE FILE (LOCATION/CONSTRUCTION)

The core file carries all the data relating to the location and construction of the borehole. Location data are carried in various forms, not mathematically related to one another, viz. co-ordinates (which are compatible with machine use), land titles (to facilitate ad hoc inquiries and linking with licensing information), and river basins (to facilitate linking with future surface water data systems). Construction data include details of all casing and screens, aquifers penetrated (depth, thickness, yield, salinities), and any gravel pack used. These are treated as series of repetitive sub-records to save space. In order to fulfil its function as the basic enquiry file, the core file also carries brief details (if available) of pumping tests, hydrogeological parameters, water salinity, any reconditioning or refitting of the borehole, present and proposed use, and other administrative data.

'Free format' notes are allowed in the core file. Such use is essential for completeness and flexibility, and does not necessarily make the data inaccessible to machine treatment. (Certain types of note are indicated by a code, and information is held within them in a fixed sequence.)

LITHOLOGY FILE

The lithology file carries drillers' logs and, where they are available, geologists' logs and grain-size analyses. The logs are reduced to a complex series of numeric and alphabetic codes for storage and manipulation, but are expanded to a more readily understood mnemonic form on the printed listings. (The lithology file and core file are combined in a single printed listing to simulate the earlier manual system.) Because of the relative brevity of drillers' logs, it has been possible to code them at the same rate at which coding of other data proceeds. A lengthy dictionary has been developed, to cover the wide variety of terms used. It is intended that original data be stored in the system, so interpretation of drillers' logs is avoided if at all possible; for example, no attempt is made to alter a log which describes alternating layers of gravel and granite. Coding of geologists' logs has turned out to be complex and time consuming because of the detail involved, and hence only an abbreviated version in the form of a drillers' log is coded at present.

ARTESIAN FILE

About 20% of New South Wales is underlain by the Great Artesian Basin. Early in this century it was recognised that water from the basin formed a major resource, and, as such, required regular monitoring. Every borehole which still has a hydrostatic head above local ground level is visited on a regular annual to triennial cycle, and a number of flow and pressure tests are carried out. This data is carried on the 'artesian flow' file, together with relevant administrative data such as the condition of the well-head and the method of delivery of the water to the supply points.

HYDROGRAPHIC AND CHEMISTRY FILES

The 'hydrographic' and 'chemistry' files carry periodic water level measurements and water analyses respectively. In both cases, the field (or laboratory) report sheet is used directly as the input document for key-punching. In general, water levels are not measured more than once a month, but the hydrographic file is capable of accepting data taken at one minute intervals if necessary. The chemistry file carries information in records relating respectively to the major ions (Ca, Mg, Na, K, HCO_3, CO_3, SO_4, Cl) and to groups of other less commonly measured determinands, including some bacteriological data. Initially the chemistry file, like the others in the system, related solely to groundwater, but it has now been expanded to include surface water analyses as well. A feature of this file is that factors such as hardness, and sodium absorption ratio, previously calculated by laboratory staff, are now calculated by the computer. In addition, the output is noted if the major ions are significantly out of balance, the current break point being an imbalance of 5%.

Some free format fields are available within the chemistry file, and are designed to accept the name and concentration of determinands not included in the fixed record format. Machine access to data in these fields can be made available by using a coding convention, such as that recommended by the Australian Water Resources Council (Anon. 1976). These fields can also be used for comments or other data, if required.

LICENSING SUB-SYSTEM

The licensing sub-system can be linked to the central system, but is essentially independent. The structure is basically similar to that of the central system, in that the borehole number and coarse location data (map number) are carried for every borehole recorded, but they do not constitute identification fields on this file. Because the licensing sub-system is oriented to an auditing and administrative function, data are held on the file in order of license number as the prime access key. The licensing sub-system is designed to issue the annual groundwater usage return forms, and reminders if necessary, and to collate the resulting incoming data, providing part of the information needed for successful long-term management of regional groundwater resources. In addition licenses due, or overdue, for renewal are listed at frequent intervals so that appropriate action can be taken.

3. DATA AND DATA ENTRY

The importance of entering accurate data into any system cannot be stressed too highly. It is difficult and expensive to track down and remove errors once they are incorporated on a master file. The quality and accuracy of the output is largely dependent on the input which must be accurate if the credibility of the system is to be maintained. Users will generally expect a far higher standard of accuracy from a computer than from a person retrieving the same data using a manual system, even though the origin of the data is the same.

Data errors can be classified as 'source errors' which originate in the measurement of the natural 'event' and the recording of this measurement on the source document, and 'transcription errors' which can arise at each stage of transcription between this document and entry on the master file.

The most obvious methods of detection of errors are by machine, either at the time of key-entry or of transaction file checking. At both stages it is possible to detect replacement of alphabetic and numeric characters, and other low-level logic errors, and when the transaction file is checked prior to updating the master file quite complex logical errors can be removed, e.g. the maximum string length of casing can be checked against the depth of the borehole and the length of screen of the same diameter used. However, many errors cannot be found logically, and those that are found are often relatively expensive to correct because of the need to return to the source documents.

In order to by-pass the accumulated transcription errors of later ledger-type and card systems it was decided to code from original documents, which meant finding files opened as early as the 1880's. This had advantages in that additional information was found such as multiple chemical analyses spread over a number of years, and, because the original files generate greater interest than a monotonous card system, the boredom of an otherwise tedious job was reduced, leading to fewer transcription coding errors. This inevitably resulted in a much longer time being required for coding all the historical data.

Transcription errors are better prevented than cured. At the coding level this means that:

(i)   some knowledge of the subject is desirable;

(ii)   the coding tasks must be planned so as to minimise boredom;

(iii) coders must not be made to feed foolish for asking advice.

It has been found that people with little knowledge of hydrogeology can, with suitable training and supervision, perform the coding task very well. On the other hand a coder with a degree in geology and several years field experience has also performed excellently. Limited periods during which one type of data is coded, help to prevent boredom. Coders are issued with a detailed coding manual, dealing with each coding field, and updated appropriately when problems are experienced. Probably the best method of maintaining a low error rate is to encourage involvement and to assign responsibility for small tasks to individual coders.

At the Commission all coded data is checked through item by item by a second coder, and apparent discrepancies discussed, before the data is passed for key-entry. There is no fixed relationship of coders and checkers, all data coding staff checking each others work at random, and nobody's work is regarded as sacrosanct. Where the coders cannot solve a problem on their own they have recourse to the Systems Officer or a hydrogeologist, but not to the programming staff. The coders also check through those coding sheets filled in outside the Section and used as source documents, (e.g. the chemical analysis forms) looking for errors in coding practice, illegibility and obvious illogicalities. Finally the transaction file is listed in its entirety whilst passing through the pre-update error detection program, and this listing is proof-read against the input coding sheets.

An important aspect of data accuracy is standardisation. It is the job of the Senior Data Coder to ensure that similar natural 'events' are always coded in the same way. This ensures that accurate use and interpretation can be made from similar items of data which may have been coded five years apart.

The design of coding sheets is very important and has a bearing on the accuracy of both the coding and the key-entry of the data. Large arrays of identical fields have been avoided as far as possible, especially where some of the fields are liable to be left blank. Coding and keying should flow, and the form should be designed to assist this.

A recent innovation has been the preparation by the computer of the coding sheets for recording water levels in boreholes in the field. The forms are issued with the identification field values pre-printed, which greatly reduces the chances of transcription errors in these fields, although slightly increasing the possibility of the entry of an identification and variable data set which are not related.

When coding and checking has been completed input documents are batched with a header sheet showing the range of documents in the batch, the dates of key-entry and verification, and the number of the cassette on which the data has been keyed. Batch header sheets greatly facilitate the location of documents for correction or re-entry where this is necessary.

The system originally accepted data on 80-column punched card, but data entry has been via cassette tape since 1974. The cassette system offers great advantages over the use of punched cards, not the least of which is the saving in storage space of cassettes in comparison to boxes of cards. The key-entry device incorporates a programmable memory and can carry out pre-editing routines, as well as the verification available on card systems, although this editing has to be limited if the rate of input is to be maintained. Correction of data is simple and involves no wasted cards.

The use of cassette entry has also allowed a reduction in mainframe processing time because it provides a maximum record size of 216 characters. When cards were used records had to be constructed from card sets with identical identification fields, a process involving sorting, assembly and editing by a special program. Data are now run on to a half inch magnetic tape from a number of cassettes, and can be run into what was originally the second program in the update sequence.

4. DATA OUTPUT

One of the main advantages of the manual card system was that the cards could conveniently be used on one's own desk, together with other references. Computer printout is much less convenient because the particular borehole records required cannot be separated from the remainder of a bulky listing. Only two years after the first magnetic file was opened the system output had become too large to be used easily at one's desk.

Computer Output Microform (COM) has solved almost all the line-printer output problem. When paper printout is required, the final program prepares a print tape which is converted into type by an off-line high speed printer. For COM a modified form of print tape is fed to a machine which produces relatively reduced photographic images on film; no inter-mediate hard copy on paper is required.

Two standard types of COM are available, microfilm and microfiche. The former is a length of film, usually held in a closed spool, with consecutive page images next to each other and an index at one end. Data location is via labels on the spools. Microfiche, which we consider more flexible, comprises 15 × 10 cm rectangles of film on which can be carried the images of a number of print pages. Data location is by large characters at the head of the fiche, included at the time of production and specified on the original print tape, and, within a particular fiche, by a grid-coordinate system in the index carried in the last page image position (see Fig. 4). Our data is carried at 42 × reduction so a single microfiche carries the equivalent of 208 pages of printout.

Microfiche is used in all those processes where paper printout would normally be used, except when extensive annotation of the listing is necessary. For reference purposes microfiche readers are available, including collapsible versions which can be used in the field. Paper copy of single pages, on which notes can be made, can be obtained in a matter of seconds from reader/printers. (The relevant data having been found, a button is pressed and a full size photocopy is ejected from the machine.)

Microfiche represents a volume saving of more than 98% compared with the equivalent paper form, and the cost of postage is nominal when transferring even quite large volumes of data to district offices. In addition microfiche originals cost only about 1/6th of the equivalent paper form, and copies are cheap (20c(12p) per microfiche), easily obtainable and invariably as legible as the original. Because of its low volume and good long-term storage characteristics, microfiche can be retained almost indefinitely and thus provide a physical back-up in the unlikely event of destruction of a master file.

It is imperative that a computer system be capable of providing data in forms other than printout. This represents perhaps the most important advance over the manual system. Data can be stripped from any file or combination of files and written to magnetic tape for analysis, plotting, or transfer to other organisations. Service programs are available to transcribe the Honeywell format tape to a format capable of recognition by any other computer available to the Commission (e.g. the in-house PDP-8/E mini-computer).

185

5. DATA EXTRACTION POSSIBILITIES

Data extraction programs exist for the Chemistry and Hydrographic
files, the user nominating the geographic area or borehole numbers required
and receiving a half inch tape formatted for use on the Hydrogeological
Section's mini-computer (PDP-8/E). Such programs can obviously be extended
quite readily to extract data falling within any combination of bracketted
variables, and to re-sort that data into any required order. Several days
must be allowed for coding, punching and compiling these program modifications. Also available for modification for data extraction purposes,
should they be of any benefit, are the many programs which are used for
maintenance purposes by the systems and programming staff.

It would be beneficial to reduce the time between the request for data
and its availability to the user, and to provide a simpler but more
comprehensive data extraction technique. To this end the specifications for
a pre-compiler for the hydrographic file have been written. This program
would allow the user to specify the basis of his search (including
comparison between individual records), and the form (data fields to be
included) and structure of the output. Input to the pre-compiler is
structured but simple, and as near to spoken English as possible, so that
users, once familiar with its use, would be virtually independent of the
programming staff. The pre-compiler will use a set of stored sub-routines
to 'write' an extraction program based on the user's requirements, and
then to compile and run it.

The hydrographic file was chosen because it has the simplest internal
structure. Nevertheless substantial problems are anticipated in writing
the pre-compiler and making it work. Once it is running properly it will
be a relatively simple process to write similar programs for other files
and file combinations. It must be recognised that such programs can do
nothing that cannot be programmed manually. They are a device by which the
user is put into closer contact with the storage system, and the programmer's
time is made available for more interesting and complex jobs.

6. PRESENT AND FUTURE USE

The implementation of any system is a slow process, and this one is no
exception. It has taken nearly six years to complete the programs which
form the basis of the system, including the time required to transfer the
system from one mainframe to another when the ADP Bureau purchased a new
computer. It is not expected that coding of the pre-1970 data will be
completed until 1985, but new data is entered on receipt so the back-log is
not growing.

Apart from the chemistry and hydrographic data extraction programs
already mentioned, present use of the system is confined to what is available

as printout, and the output from a number of management oriented programs.
Nevertheless substantial advantages are already apparent because of the
mode of storage. The licensing sub-system, for instance, holds the most
comprehensive set of licensing data available within the Commission, and
is regularly used to obtain information which was previously inaccessible.
The system provides management with up-to-date lists of observation
network boreholes, and routine chemical analyses and water level observations in these networks, enabling rapid identification of areas where
data-collection is redundant or has been omitted. A very great advantage
has been the systematisation of data collection by individual geologists,
and the release of large volumes of project oriented data previously hidden
away in filing cabinets.

The system has the potential for providing regional groundwater
contour maps, delineating groundwater and surface water quality anomalies,
modifying observation networks in space and time to suit changing requirements and improve efficiency, and feeding calibration data directly into
numerical resource models. When the modular design of the system is taken
into account, and the resulting ability to add new files and sub-systems
at will, the long term possibilities will be seen to be very great indeed.

7. CONCLUSIONS

For the data storage and retrieval system in use at the Water Resources
Commission to provide maximum benefit, it must eventually contain all the
hydrogeological data available. Even in the present incomplete state it is
assisting in the day to day management of office and field services, and
has made data previously inaccessible, and even unknown, available to all
users. Its presence has led to a systematisation of collection procedures
which has brought further benefits outside the sphere of computing.

The coding of data from the original documents has reduced the speed
of data entry considerably, but has removed several significant sources of
error. Much of the implementation effort has gone into the prevention,
and cure, of the entry of erroneous data. The main form of output is
microfiche, which has considerable advantages over paper printout in terms
of storage, ease of use, availability and legibility of copies, and cost.

Because the data are in machine compatible form, types of processing
are available which could not have been contemplated with a manual system.
The data can, for instance, be re-sorted and presented in any required
order or can be extracted and analysed according to any combination of
constraints.

The groundwork of the system is complete, and data extraction and
output in any form is potentially available. The future of the system lies
in the development of programs which will automate the current conventional

uses of it and release the programming staff to deal with more complex problems.

ACKNOWLEDGEMENTS

The authors are grateful to their current employers, the Water Resources Commission of New South Wales and the Water Research Centre, for permission to present this paper. The project is entirely funded by the Water Resources Commission, and all the work entailed is in the hands of staff of that body. Staff of the State Government's ADP Bureau have made the project possible with their consistently patient and invaluable advice.

REFERENCES

ANON., (1976). Standards for interchange of water resources data on computer media. Australian Water Resources Council. Hydrological Series No. 10.

LORIMER, J., WOOLLEY, D.R. and LOWE, I., (1970). Management report on the storage of bore data. Water Resources Commission of NSW (Unpubl.).

NEW SOUTH WALES GOVT., (1976). Water Resources Commission Act, 1976. NSW Govt. Printer, Sydney.

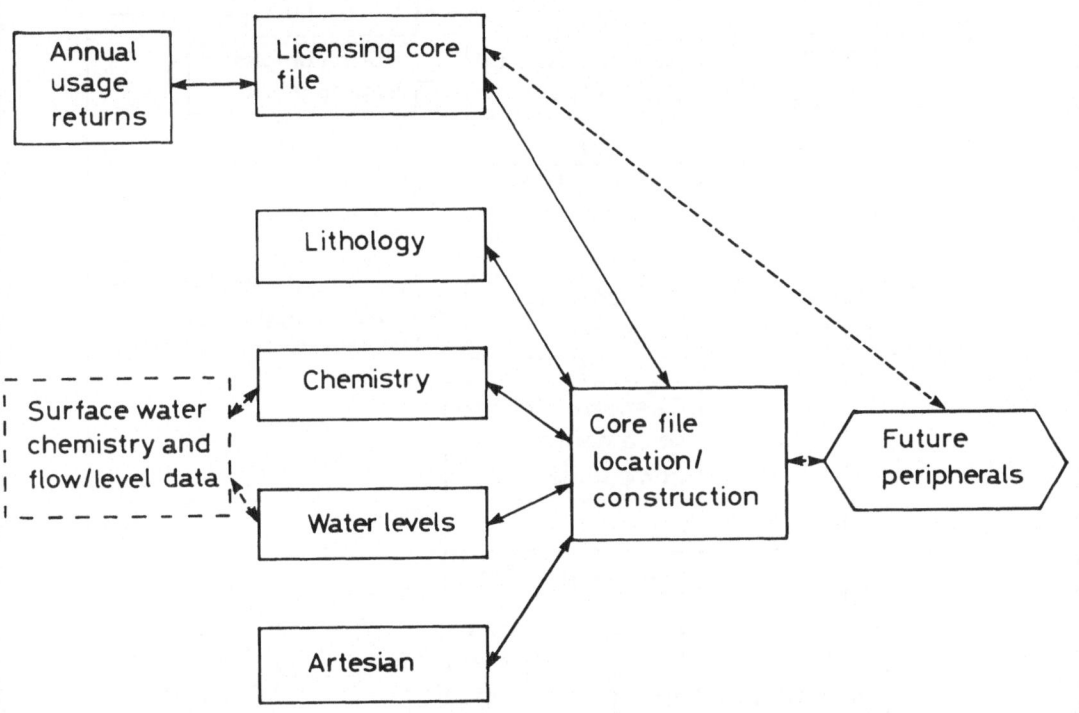

Figure 1.   The system structure.

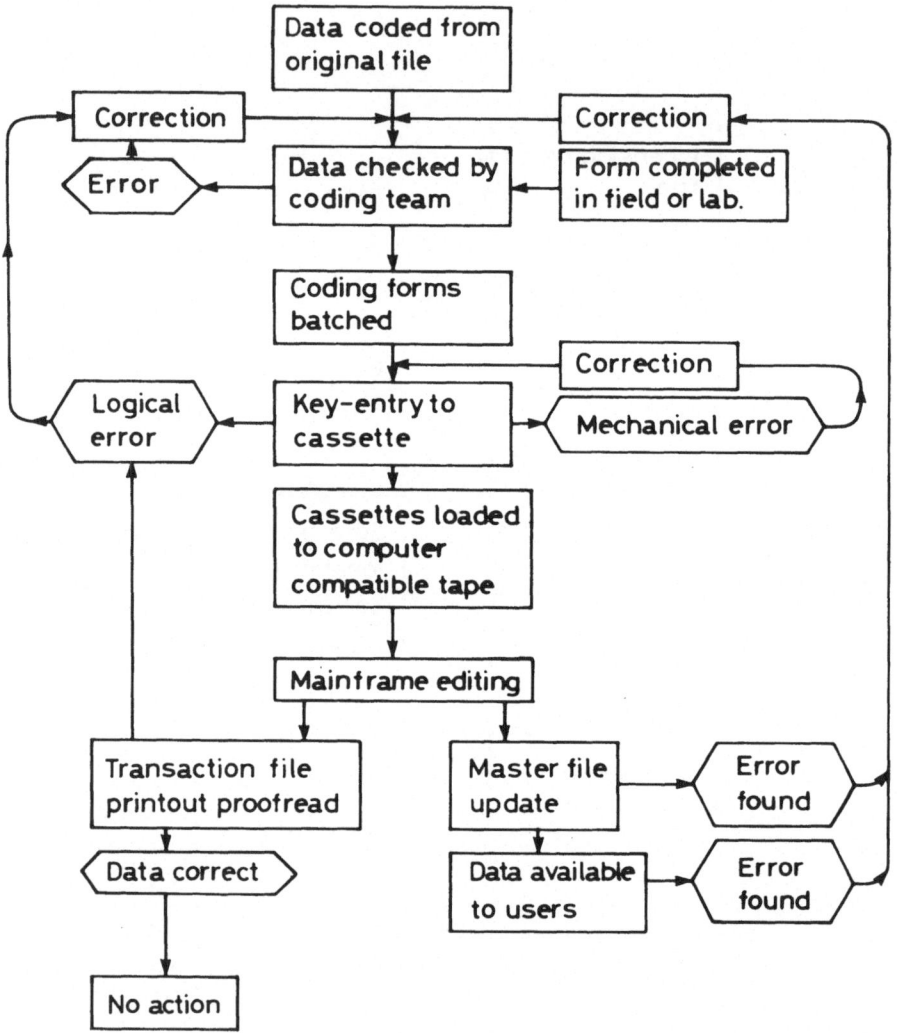

Figure 2. Simplified data entry and error detection flow path.

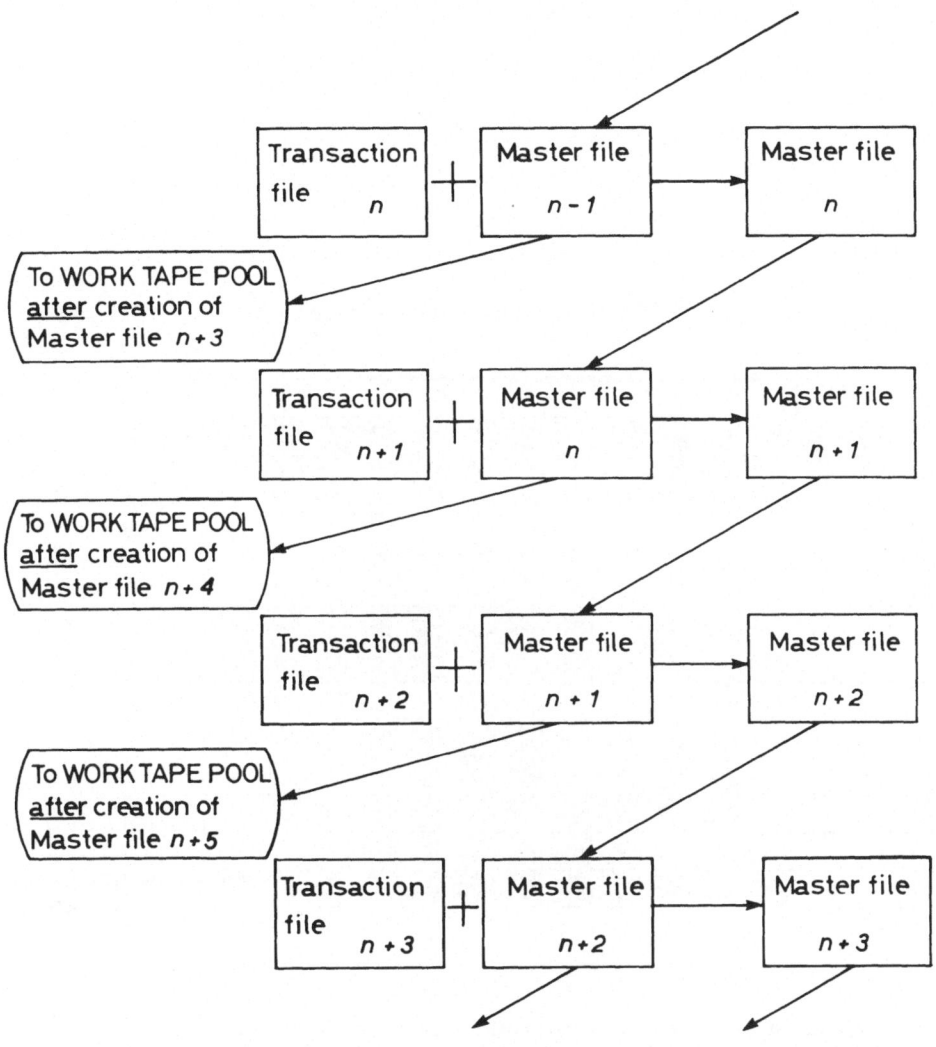

At any given time the master file series comprises 4 extant files:

$n-1$ great grandfather (released to work tape pool after next update)
n grandfather (becomes great grandfather after next update)
$n+1$ father (becomes grandfather after next update)
$n+2$ son
$n+3$ to be written during next updating run.

Figure 3.   The 'grandfather/father/son' updating system.

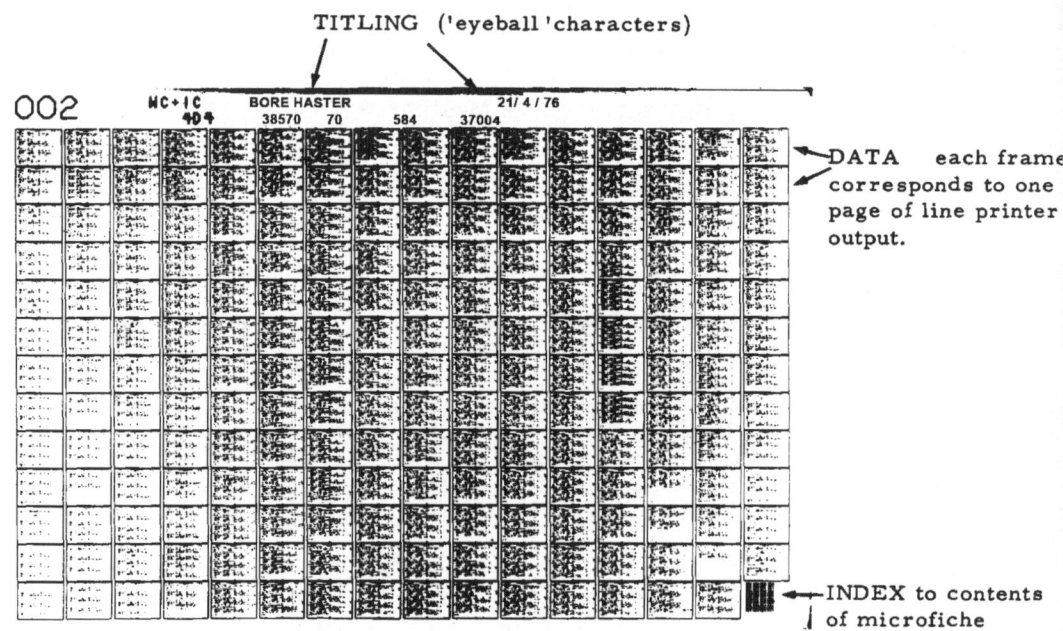

Figure 4.   Layout of a typical COM microfiche from the system.

APPLICATIONS, AVAILABILITY AND ACQUISITION

OF EARTH SATELLITE IMAGERY FOR GEOLOGY.

NIGEL PRESS

Earthsat Data Centre

Nigel Press Associates Ltd, Apex House, Edenbridge, Kent.

Summary: The degree of perspective given by the wide field of
view of satellite imagery has opened up a new panorama for geological studies in recent years, particularly in the field of
structural geology. Additionally, in areas where aerial photo
graphs cannot be obtained, satellite photographs will almost
certainly be available, and may form a valuable substitute.
For the Librarian, however, there are particular problems in
acquiring and handling satellite imagery, which hopefully are
clarified by this description.

1. GEOLOGICAL APPLICATIONS OF SATELLITE IMAGERY

The geological applications of satellite imagery can be summa-

rised as follows;

i) very graphic and readily updated base map; coverage at

different times of the year to show seasonal changes is pos-

sible. Scales of upto 1:25,000 can be reached in some cases,

although 1:250,000 is more normal. No radial distortion.

ii) Source of geological, lithological and structural, in-

formation; few geologists will fail to percieve this on stu-

dying the imagery, although some areas are more amenable to

interpretation than others.

iii) Since the imagery is available in digital form, it can

be fed into various other data bases quite easily, and subtle

and complex manipulations and enhancements are possible.

Computer time, software and plotting facilities can be expen-

sive however.

iv) Satellite imagery is a quite unique source of tectonic

information on a scale which is normally only obtainable with

geophysical surveys. Although the full implications of linea-

ment patterns have yet to be definitively evaluated, they will

undoubtedly lead to new and better understanding of global

tectonics.

2. AVAILABILITY OF SATELLITE IMAGERY

The sources of imagery can be classified as follows;
1) Manned Satellites
2) Unmanned Earth Observation Satellites
3) Unmanned Meteorological Satellites

2.1) MANNED SATELLITES

Imagery from manned satellites such as the Gemini and Apollo series used in the American Space Programme of the 1960's generated the first useful pictures of Earth for Space, and aroused geological interest in this type of information. Although Gemini and Apollo imagery is still used from time to time, most of the shots are obliques and therefore of illustrative rather than interpretive value. Some of the later Apollo missions carried vertically oriented cameras for earth observation applications. Better quality imagery was obtained during the Skylab missions of 197314 when both oblique and vertical photographs were acquired, including multispectral photography (S190A) and photographs from a high resolution metric terrain camera (S190B). The Soviets have also acquired similar photography on their Soyuz Missions, and considerable new information will be acquired by the joint American/European Spacelab in the 1980's.

The principal limitations on the use of imagery from manned satellites is the restriction on availability of ground coverage due to;
a) the equational orbits, rarely exceeding 35° of Latitude
b) highly limited number of photographs possible in each mission.

At a 'guesstimate' between 10 and 20% of the Earth's land surface is covered, although the United States is well covered. A wide variety of different films and filters have been used, and in some cases stereo coverage is available.

2.2) UNMANNED EARTH OBSERVATION SATELLITES

The Landsat (formerly called ERTS) satellites are the main members of this group, and provide by far the greatest source of imagery for geological applications.

The satellites are in a polar orbit at an altitude of 980kms
and have a ground imaging swath 185kms wide. The ground track
is repeated every 18 days, so theoretically imagery is available repetively, although this is rarely the case in practice
due either to cloud cover or lack of data transmission and recording facilities. However, cloud free imagery of approximately 95% of the Earth's land surface has been aquired by the
multispectral sensors onboard the Landsats which have been in
operation since 1972. The latest satellite, Landsat 3 has a
thermal infra red channel which produced some imagery before
failing, and also a Return Beam Vidicon (RBV) Camera with
40m. resolution (i.e: double the resolution of the MSS);
limited coverage of this imagery is currently being acquired.

The principal advantages of the Landsat system are;
1)   polar orbit allows acquisition of globally uniform cover.
2)   repetitive coverage is possible to monitor seasonal vari-
     ations.
3)   the multispectral sensor allows separation of geological
     units or geobotanical associations on the basis of spectral
     reflectance properties.
4)   the data is available in digital form, so can be readily
     manipulated in the computer.
One of the main restrictions on the Landsat system is the lack
of penetration of cloud cover. However, a recent experimental
satellite,Seasat had an active imaging radar system which is
not affected by cloud cover. Several thousand line-kilometres
of this imagery was acquired over land areas before the satellite failed, and this will eventually be available for general evaluation by geologists.

Future planned unmanned satellite systems include higher resolution multi-spectral sensors, active radars, thermal infra-red
sensors and stereo-imaging systems, all of which will be valuable in geological studies.

2.3)   METEROLOGICAL SATELLITES
The only imagery from current satellites likely to be of use is
from the VHRR sensor on the polar orbiting NOAA satellite.
The resolution of the imagery is approximately 1km,but this

may be useful to show up major tectonic features which could only otherwise be found on mosaics of many Landsat images. This can be particularly important for areas with extremely poor weather conditions such as Alaska, Greenland, Antarctica etc, where cloud free conditions rarely coincide with Landsat passes, but NOAA overpasses occur daily.

3) ACQUISITION OF SATELLITE IMAGERY

3.1 INDEXING AND CATALOGUES

Almost all the satellite imagery referred to here is acquired by the United States Government Agencies and made available in the public domain by them. The principal source is the EROS Data Centre, Sioux Falls, South Dakota 57918 U.S.A., except for Meteorological Satellite Data which is available from S.D.S.B./N.C.C. World Weather Building Washington D.C. 20233 U.S.A. Any interested users are recommended to write for information, but enquiries should precisely state the geographic co-ordinates of the area of interest.

The EROS Data Centre has been set up by the U.S.G.S. specifically by disseminate Earth Resources Satellite Data to users and guide them in their application. Catalogues of the imagery are available. The Gemini and Apollo catalogues are not of great use and cannot easily be searched nor are there good ground cover maps. Excellent catalogues and ground cover maps are available for the Skylab missions. Landsat imagery is a somewhat different problem, since with over half a million images archived, hard copy catalogues are very bulky. At present the catalogues are being converted to microfiche, but the data is all stored in computer files, so specific computer searches are normally the best way to search out available imagery.

Landsat imagery is indexed in three ways;

a) by an image i/d number specific in time to the moment of acquisition (from which geographic location can be calculated).

b) By geographic co-ordinates of image centre point or corners.

c) By a numbering system of satellite paths and down path row. (Global path and row indices can be obtained from EROS D.C.).

Other parameters which have to be taken into account are image
quality and cloud cover.

The World Bank has produced a Landsat Image Cover Atlas for the
Developing Countries of the World (available from World Bank,
Cartography Section 1818 H St. NW Washington, D.C. 20433 U.S.A.)
For most purposes the optimum way to obtain information is to
request a computer search from the EROS Data Centre.

3.2   FORMATS FOR ACQUISITION

Most of the imagery from unmanned satellites is available as
70mm format negatives or positives which can be handled by
almost any photographic laboratory. However for Skylab S190B,
Meteorological Satellite imagery and future Landsat products,
the format is 23cm. square which can normally only be handled
by more specialized photo-labs. Although Landsat imagery will
be available in 70mm format for some time to come, there is
the additional problem of making 'colour composite' images from
the multi-spectral data which requires 23cm. format imagery.
Colour composite generation is quite a complex procedure and
should be evaluated carefully before it is attempted (Kreitzer, 1974).   If required,black and white or colour prints can
be ordered, but the problem here is lack of choice of format/
scale, control over density, contrast or colour balance, and
lack of flexibility for future reproduction.

3.3   LANDSAT DIGITAL DATA

Although most geologists have seen the visually striking pictures produced from satellites, few are aware of the computer
data processing which has to be undertaken to create these
pictures or of the enormous amount of further information
which can be extracted from the raw data by further digital
processing. Raw Landsat is available in the form of a 'computer compatible tape' (C.C.T.) normally as a 1600b.p.i. 9
track tape. In this form the data can be subjected to far
more detailed analysis, but procedures are more complex and
costly than with photographic imagery.

The main sensor on board Landsat is a multi-spectral scanner
meaning radiance in four discrete spectral bands from pixels
(picture elements) with a ground area of 79m x 56m Approx
3200 pixels are recorded along each scan, and the scans are

recorded as continuous strips later formatted to images 2300
scans deep i.e: approximately 7 million pixels per band.
Each pixel is an 8 bit byte with a dynamic range of radiance
from 0-255.
From the forgoing figures it will be clear that a fairly major
data processing effort is required to produce significant
image enhancements such as spectral band ratios or thematic
signature extractions..

Not only does special software have to be developed (and in
the case of interactive systems with colour T.V. display,
special hardware), but plotting devices for hard copy output
are both expensive and difficult to find.

4. CONCLUSIONS
Hopefully the foregoing points will not by their complexity
have deterred potential users from acquiring satellite imagery.
At least they should point out some of the methods of acquiring the data, and also some of the pitfalls to be avoided.

5. REFERENCES

GENERAL

Skylab Earth Resources Data Catalogue NASA Publication
JSC 09016.

ERTS-1, A New Window On Our Planet. U.S.G.S. Prof. paper 929.

Mission to Earth, Landsat Views the World
Nasa Special Publication SP 360

Kreitzer, 1974. Direct Additive Printing. Photogrammetric
Engineering pp. 281-285.

BIBLIOGRAPHICAL CONTROL OF GEOLOGICAL MAPS

JUDITH A. DIMENT

British Museum (Natural History), London SW7 5BD

AND

JOHN R. SCHROEDER

Library of Congress, Geography and Map Division, Washington DC

Summary: A geological map is defined and the history of geological maps briefly outlined from the 18th century to the present day. The current status of geological mapping and the problems of acquiring geological maps are reviewed including the lack of adequate bibliographical tools, and the nature and form of map publication. The problem of maps published in serials and monographs is highlighted. It is suggested that the most serious problem of bibliographical control is the lack of adequate bibliographical details on the maps themselves and a call is made for improved bibliographical standards at the publishing stage. The problems of cataloguing and classification are outlined stressing the benefits and requirements of cooperative cataloguing systems such as the Ohio College Library Center; the standardisation of map classification and the importance of the Library of Congress system; the standardisation of descriptive cataloguing and the improvement and standardisation of subject access. The extension of cataloguing coverage to include geological maps in monographs and serials is discussed and means of improving network capabilities are suggested.

Introduction

1 History of geological mapping

2 Status of geological mapping

3 Status of bibliographical control

    a Map format

    b Acquisition problems

    c Automated programmes

4 Cataloguing and Classification

    Progress towards bibliographical control of geological maps

    a. Library systems development

    b. Cooperative cataloguing benefits and requirements

    c. Special cooperation among geological libraries

    d. Effective use of existing network capabilities

    e. Standardisation of map classification

    f. Standardisation of descriptive cataloguing

    g. Improvement and standardisation of subject access

    h. Extension of cataloguing coverage to include geological maps in monographs and serials

    i. Improvement of network capabilities

Introduction

"One of the more annoying things about maps is that it is often difficult to know if one exists for your area and theme of interest and at a useful scale ..... finding out requires patience, skill and an excellent local library. Matters get worse when you find there is such a map but it is not available locally." (Rhind, 1977) This heartfelt user's cry reflects many of the problems inherent in any form of bibliographical control and not least in the bibliographical control of geological maps.

A geological map provides a record of deposits and structure of a region, including details such as faults, mineral veins and fossil localities. They reflect the extent and accuracy of geological knowledge at the time of preparation and are an important method of recording geological information and discoveries. They form a basis for current and future research and have been described by North (1928) as a dynamic force in geology. Geological maps are therefore in frequent need of revision. As Dunham (1967) has pointed out, a geological map is always in some degree an interpretation and not merely a record of fact, and revision and reinterpretation are necessary in the light of new evidence.

History of Geological Mapping

The history of geological maps consequently reflects the history of geology itself e.g. there is an interesting contrast in many of the maps produced in the early nineteenth century depending on whether the Wernerian or the Huttonian school was followed. The earliest geological maps issued in the second half of the eighteenth century showed mineral deposits e.g. those of Guettard (1746, 1752) in the Memoires of the Academy of Sciences, Paris which include mineralogical maps of Switzerland, France and England. These early maps were very much the work of individuals as were the first 'proper' geological maps produced in the early nineteenth century by such people as William Smith, William Maclure and Leopold von Buch. The first major landmark in geological mapping must surely be William Smith's hand coloured map of England and Wales, published in 1815, as this is the first attempt to show the geology of a whole country on a single map.

The difficulties encountered by individuals trying to map large areas were many and varied and this inevitably led to the formation of national geological surveys such as the Geological Survey of Great Britain founded in 1835. But it was not until the second half of the nineteenth century that geological mapping was established in many countries throughout the world. The development in printing technology in the latter half of the nineteenth century accelerated this development and by the end of the century many countries were producing large scale geological maps.

The first circular announcing the first International Geological Congress in Paris, 1878, issued in 1876, made a call for an exhibition of geological maps and sections to illustrate the laws of mountain structure.
(International Geological Congress, 1876) It also stated that discussions should be held on such topics as scale, colours and symbols and other methods of representation on geological maps to prepare the way for improved geological maps of continents.

At the second Congress in Bologna in 1881 a Commission for the International Geological Map of Europe was constituted. (International Geological Congress 1881). This map was published in 1893 and is the first of a series of small scale international maps. At the twelfth Congress in Canada in 1913 a proposal was made to produce a new geological map of the world. (International Geological Congress, 1913). Out of this proposal grew the Commission for the Geological Map of the World, which today plays such an important part in publishing small scale geological maps of continents including the Geological World Atlas 1:5 000 000 (Unesco, 1977). The maps produced by the various Commissions during the twentieth century reflect the growth in cooperation between surveys and mapping agencies.

Increased specialisation of geology in the twentieth century are reflected in the maps produced, including tectonic maps, metallogenic maps, glacial maps, stratigraphic maps, hydrogeological maps, geophysical maps, geomorphological maps, oceanographic maps and more recently extraterrestrial maps and the ERTS satellite imagery.

Status of Geological Mapping

The status of geological mapping throughout the world was reviewed by Dunham (1967) who points out that only a very small proportion of the earth's surface has been mapped on scales larger than 1:100 000. If the 1:63,360 scale is considered as the minimum necessary for the detailed recording of individual exposures, then this level of mapping has been achieved in very few areas. Although there are world maps, such as the Geological World Atlas (Unesco, 1977) these give only the very broad outlines.

Dunham (1967) produced an index map of world geological maps

published between the scales 1:25 000 and 1:1 000 000. This map
highlights the patchy nature of geological mapping, reflecting in part
the political history and the programmes of foreign aid at present.
Present geological mapping projects can vary from large-scale systematic mapping of a whole country by an official body; a specialized
mapping project of a small part of a country, often for specific
economic reasons (e.g. a small area in Pakistan was mapped by a team
of Canadian geologists as part of the Colombo project (Jones 1961));
or purely for academic research. A knowledge of the funding of
mapping projects is important because it can affect where the map is
published e.g. a map of Guatemala was published recently by the
Bundesanstalt fur Bodenforschung (1971)

In 1978, it was estimated that about 2000 geological maps, at
various scales, are published annually in the world, by three major
groups of publishers:
1. National geological surveys and other governmental bodies (this
   is the most important group and they are usually responsible
   for large-scale detailed mapping).
2. Commercial map publishers
3. Scientific societies and other learned bodies

Status of Bibliographical Control
a. Map format
The map format has received inadequate and unequal levels of bibliographic control in both general and geoscience libraries. The primary
reason for this inadequate control seems to be "..... that
librarians have too little understanding of maps to give them the
attention they deserve as sources of information ..... (Erlach, 1961).
Because librarians have not recognised the research value of the map
format they have not recognized their "responsibility to acquire,
control and provide access ..... to information irrespective of format"
(Daehn, 1975)

Libraries are service institutions, which throughout their
history have had problems in obtaining adequate funding. As book -
oriented institutions libraries naturally enough have given the book
priority in the allocation of their financial resources. In addition
maps have inherent physical and bibliographic characteristics, which
have made them a long-standing problem for such book-oriented
librarians. Maps are difficult to acquire, expensive to store,
maintain and preserve. They are cumbersome to retrieve, circulate

and refile. Maps are issued in various formats (e.g. flat, folded, rolled or as atlases, and globes) as well as being issued in books and journals and on computer magnetic tape.

b. Acquisition problems

The very diverse nature of maps has resulted in problems of acquisition as very few booksellers stock maps. Of course there are specialised map sellers such as Geocenter and Telberg but even these do not carry comprehensive stocks.

One of the main problem in the bibliographical control of geological maps is the lack of adequate bibliographical tools. At present there are none designed specifically for the control of geological maps. General bibliographical tools such as Winch's International Catalogue or Maps in Print (1976) can be used but works of this type only include the major geological maps in print. To keep abreast of new maps it is essential to search publishers' catalogues, accessions lists, bibliographies, annual reports of surveys, geological, geographical and cartographical journals. National bibliographies are another area where maps have been neglected but events in recent years look encouraging, for example, the French and Italian national bibliographies include maps. The first Dutch national map catalogue was published recently. (Bibliografie van in Nederland, 1975). Discussions are in progress concerning the possibility of including maps in the British National Bibliography. For many years geological mapping in the United States has been reasonably well covered in publishers' catalogues, published index maps including the Geological Survey publications, catalogues of the surveys of individual states, and also in the Bibliography and Index of Geology.

Another difficult problem is that of linking the maps with their explanatory texts as they are not necessarily issued at the same time. It is often therefore essential to scan the primary geological literature to link the maps with their texts eg. the Geological Map of the Arabian Peninsula (1963), the accompanying text was issued as Professional Paper United States Geological Survey number 560. This problem is made even more difficult if the map collection is divorced from the library.

Keeping track of map serials also has many problems as the same series may be issued in different guises - it may be issued flat with or without a separate explanation, or folded with or without an explanation or issued as part of another series altogether e.g. 'A'

series of Geological Survey of Canada which is sometimes issued flat
as a separate series and sometimes issued folded in the Paper series.
Another example is the United States Geological Survey Miscellaneous
Investigations Series : these need careful scrutiny as they include,
for example, the Liberian Geological and Geophysical series as well
as the Geologic Atlas of the Moon and Geologic Atlas of Mars series.
There is a need for a comprehensive list of geological map serials,
both current and retrospective.

An even greater problem area are the geological maps that are
published in journals and books. They range from very detailed maps
of very small areas illustrating a new piece of research to maps such
as the Geology of Donegal issued in a book by Pitcher & Spencer (1972)
and the Geology of the Pacific Ocean published with the Deep Sea
Drilling Reports (Heezen & Fornari 1975). These are a very important
group of maps which lack any form of bibliographical control; they
are rarely catalogued by libraries or included in secondary services.
One interesting project to deal with this sort of material was
established by the National Library of Canada and the Université de
Laval. It is called Cartomatique and is an information system which
aims to gather these maps on coded microfilm. Unfortunately, indexing
for Cartomatique ceased in the summer of 1977.

Perhaps the most serious problem of bibliographical control of
geological maps is the lack of adequate bibliographical details
on the maps themselves as this creates numerous problems of
identification. The problem is magnified when the maps are listed
in publishers' catalogues etc. It is often difficult to identify
accurately a specific map from such a list because vital information
is lacking e.g. the Geocenter catalogue (Geokatalog 1978) omits
the date. This means it is impossible to identify different editions
of one map. Variation in the title may also occur, especially if the
map is in more than one language. The recent British Standard for
the bibliographical description of maps in accessions lists and
periodicals (British Standard Institution, 1975) should help alleviate
this problem.
One of the main requirements for improving bibliographical control
is the establishment of an internationally agreed standard for the
provision of bibliographic data on each map. Such a course of action
is already advocated by the Bundesanstalt fur Bodenforschung, Hannover
which includes the following strip on all the maps it issues:

Geologische Karte von Niedersachsen 1:25,000	B1.4227 Osterode am Harz	Hannover 1976.

Every map published should state clearly the title, author, publisher,
bodies responsible for the mapping, scale, date of mapping and date
of publication, edition, number of sheets and where relevant the name
of the series. If bibliographical standards are improved at the
publishing stage this should lead to better catalogues, accessions
lists and bibliographies and should alleviate many of the problems
of the acquisition process.

c. Automated programmes

There are four major automated programmes specifically concerned
with controlling geological maps. The US Geological Survey's Geologic
Index Maps system contains extensive bibliographic data and permits
on-line access to new information as well as the retrieval of selected
data and its reconstruction into various combinations of text and
graphics. The cartographic material covered by the system is currently
restricted to geological maps covering parts of the United States.
However the system could be used for other areas and other subjects
(Fulton & McIntosh, 1976)

Geosystem's 'Geoarchive' bibliographic data base includes
geological maps in the map collection of the Institute of Geological
Sciences, London. 'Geoarchive' can be used to produce graphic indexes,
retrieve by geographic coordinates and access topical subjects,
geographical areas and geologic time on a partially pre-coordinated
basis. The 'Geoarchive' data base is currently available to users
via the Lockheed DIALOG System (Lea, 1977)

The American Geological Institute's 'Georef' system attempts
to attain worldwide coverage of the literature of geology. (Walker,
1977). 'Georef' includes citations for maps included in recent
volumes of the Bibliography and Index of Geology as well as special
map bibliographies which have been included in the system on a
selected basis. 'Georef' is available to users via Systems
Development Corporation ORBIT system.

Through participation in the Ohio College Library Center's Map
Cataloguing Sub-system the US Geological Survey Library produces
catalogue cards for cartographic items acquired in the Library.
The system provides additional service potential for US Geological
Survey researchers by means of on-line access to the Ohio College
Library Center's data base which includes geological maps held and
catalogued by other geological libraries.

However, despite the high level of bibliographic activity, the level
of bibliographic control for geological maps is not adequate to meet
the research requirements of geoscience map libraries and geoscientists. This assessment is based on the following premises
using examples from the United States.

1. The retrospective collections of the major US Government libraries
holding geological maps, including the main US Geological Survey
Library and the Library of Congress, have not been catalogued. The
contents of these collections have therefore been 'lost' to the
carto-bibliographic system and to all users who do not have direct
access to the respective collections.

2. Although some smaller geological map libraries have catalogued
their collections e.g. University of Illinois Geology Library
(Tanaglia, 1977) and the US Geological Survey Menlo Park Branch
Library, the cataloguing information is not directly available
to reference librarians or researchers in other institutions.

3. Maps, including geological maps, have never been included on a
systematic basis in the National Union Catalog produced by the
Library of Congress.

4. The coverage of the automated systems active in controlling
geological map bibliographic information is not comprehensive for
either current or older geological maps. Also, the automated
systems are not designed to include "union catalogue" information
indicating alternative locations or availability.

Cataloguing and Classification

Maps are more difficult to catalogue than books partly because
of the intrinsic problems in the process of describing a graphic
format in words, and also because of a lack of adequate cataloguing
rules for maps. Until recently the rules and guidelines for
cataloguing maps were inadequate and did not permit precise, accurate
and consistent cataloguing. There has been a lack of uniformity
in the way major research libraries have treated maps. Many
libraries have chosen not to acquire maps on a large scale or have
decided not to provide formal bibliographic control for maps in
their collection. Because libraries have not provided users with
access to cartographic information on an equal basis with books,
many have been unable to obtain information relevant to their needs.
Although map formats present libraries with unique problems in
acquisition, storage and bibliographic organisation their value is
great enough to warrant full bibliographic control on an equal basis

with other types of publications. Unfortunately the cost of full
catalogue entries for maps is going to be as high as equivalent
entries for books because the basic elements of description and requirements for authority control are directly comparable. This must be
realised by library managers if adequate bibliographic control of maps
is to be achieved in the library.

Progress towards bibliographical control of geological maps

a. Library systems development

General progress in the development of library information systems
and their adaptation and application to the requirements of maps have
made full bibliographic control feasible. Landmarks in the progress
toward the control of geological maps have been:

1. The development and implementation of the MARC Map format as an
operational system at the Library of Congress Geography and Map
Division. This is significant as it became the foundation of later
automated MARC - compatible systems. The Library of Congress
publication Data preparation manual for the conversion of map
cataloguing records to machine readable form (Carrington & Mangon,
1971) was a major development in the establishment of the MARC Map
format. The Manual represents the first published example of the
operational compatibility between the basic MARC Monographic format
and the other special MARC formats. The development of compatibility
among MARC communications formats is being continued internationally
through progress toward UNIMARC.

Parr (1975) provides a cogent explanation of the significance of
the MARC format to automated bibliographic control of cartographic
material in that it is becoming the "... internationally accepted
standard supported by major government subsidised programmes in the
United States, Canada, Great Britain, France and Germany ... and that
for map collections to be accessible through libraries generally they
must participate in MARC"

2. The implementation of the Ohio College Library Center (OCLC) Map
Cataloguing Sub-system provided map and geoscience libraries in the
United States with immediate, widespread access to a cost-effective,
on-line, cooperative map cataloguing system. Map libraries thus
achieved instant equality with other libraries in terms of having
the capability of cataloguing cooperatively. General libraries
without map collections or a map librarian gained a structured,
understandable mechanism for controlling different types of printed
information.

b. Cooperative cataloguing : benefits and requirements

The economic benefits and cost effectiveness of on-line cooperative
cataloguing have been widely reported in the literature of librarian-
ship. The allure of a system which permits sharing of the cost of
original cataloguing, direct access to Library of Congress cataloguing
records and the potential for improved reference capabilities is
understandable. Daehn (1975) has described the advantages of such
a cooperative system as:

- Shared cataloguing
- The development of a regional union catalogue
- The production of book catalogues or catalogue cards for
  individual institutions.
- The transfer of information between regional data banks and the
  national data bank.
- Improved local control of collection development.
- Inter-university borrowing of maps
- Specialised user services such as on - demand bibliographies,
  Selective dissemination of information etc.
- Provision of statistics at the local level to aid map collection
  management and planning.

He also emphasises the importance of the standardisation of cataloguing
rules, classification, subject access and machine communications format
to the success and workability of the regional cooperative cataloguing
system.

c. Special cooperation among geological libraries

Development of formal channels of cooperation and coordination
within the US geoscience library community is possible for those
libraries participating in OCLC and other similar regional networks.
This will make maximum use of their efficiency and utility for the
cataloguing and bibliographical control of geological maps. Once
the basic standards of the general system are met additional
cooperation among specialised users with common interests becomes
possible.

d. Effective use of existing network capabilities

The first objective of special cooperation among geological
libraries should be improving their own cost effectiveness in the use
of existing network capabilities. The first step in improving the
cost effectiveness of existing map cataloguing network capabilities
should be the development of channels of communication and of mutual
support among participating libraries. The next step in promoting

the effectiveness of individual institutions could be the establishment of an advisory group to counsel and inform geological map
librarians, geological library managers and user groups (such as
professional geological associations) on the implications and ramifications of a transition from informal methods of bibliographic control
of maps to a programme entailing formal, automated bibliographic
control. Increased financial support and official recognition of
the value of map collections in the library are the most important
factors to be recognised in gaining bibliographic control of the map
format.

e. Standardisation of map classification

The OCLC Map Cataloguing Sub-system allows each participating
library unlimited freedom in the selection of a classification system
for its own collection. However, because of the cost-effectiveness
and other advantages of standardised classification, geological map
libraries participating in OCLC or analogous systems should view map
classification as an area in which special cooperation could promote
effective, efficient use of the general systems. Geological
libraries participating in or changing to automated cooperative map
cataloguing programmes should consciously re-evaluate the
effectiveness of their classification system taking the following
factors into account.

1. Cost effectiveness and other advantages of classification
   standardisation including:
   Lower processing costs
   Potential for cooperative collection and acquisitions management
   Improved efficiency in administering inter-library loans
   Automated subject search capabilities

2. The specific classification requirements of an individual library
   including:
   - Physical arrangement of the collection and its subsidiary
     effect on:
     Retrieval of maps
     General collection maintenance
     Collection accessibility (open access versus closed access)
     Space, equipment or other limitations
   - User familiarity with an existing system
   - Compatibility of subject and area elements between the library's
     book classification system and its map classification system
   - Cost of conversion from an existing map classification system

to a standardized system.

- Applicability of potential standard classification systems to
  its specific functional requirements

In selecting a single, preferred standard geological library map classification system, cooperating libraries must consider the theoretical and practical classification requirements common to geological libraries as well as the future probabilities for standardisation of classification by general map libraries. In descending order the following levels of general map classification standardisation could be considered:

Universal, world-wide standardisation

By continent or groups of continents

By country

By network

Within networks

Unfortunately, it may be too late for universal standardisation of geological map classification. Although the technology for the exchange of carto-bibliographic information on an automated, international basis exists, the practical arrangements for doing so have not yet evolved. In the meantime libraries have begun entering data into existing systems without coordinating classification. Once automated programmes are established without coordination of classification it will be difficult to change to a universal classification standard.

Parr (1975 p 65) recommended that the Library of Congress classification system be used as the standard classification system for map records input into MARC format compatible systems. In North America, it is almost certain that the Library of Congress map classification will become the predominant map classification system for automated programmes. Reasons for this include: the predominance of Library of Congress book classification in research libraries in the United States and Canada; the Library of Congress 'G' schedule for maps is the system most frequently associated with the MARC Map format; the Library of Congress MARC Map data base contains over 50,000 records and the Library of Congress 'G' schedule is by far the most widely used classification system in map libraries in the United States and Canada. The only major disadvantage of the Library of Congress "G" schedule for open access geological libraries is that its inherent emphasis on sub-area over subject has the net effect of scattering subject maps

within each of four separate files under each numbered base area.
This affects the physical accessibility of the collection for purposes
of subject oriented browsing. However, browsing becomes much less
important in a catalogued, controlled collection which has subject
and area access by computer.

f. Standardisation of descriptive cataloguing

Daehn (1975) emphasized the importance of descriptive cataloguing
standards to the efficient use of cooperative map cataloguing systems.
He recommended that Anglo American Cataloguing Rules (AACR) be used
as the basis for standardized cataloguing rules on the grounds of
user familiarity, widespread use in libraries and compatibility with
catalogue records for other types of material. Consistent selection
and transcription of main entries, added entries, titles and
alternative titles are essential for accurate record retrieval on
automated systems and for listing in bibliographies. In the OCLC
Map Cataloguing Sub-System, map citations are searched under the
standard OCLC search patterns (author, author-title and title)
The search pattern retrieves added entries and alternative titles
as well as primary titles and main entries. Because of inherent
map format characteristics and because of the inadequacy of the first
edition of AACR map cataloguing rules (Hill, 1977) the capability
of searching by alternative titles and added entries is essential
for the precise retrieval of specific map records from large data
bases.

As the map cataloguing rules are in transition, consistent
description of maps for automated cooperative cataloguing systems
will be very difficult until the second edition of AACR is widely
adopted. However, access to an established AACR derived map cataloguing manual, such as the ones developed in the Library of Congress
Geography and Map Division, the Newberry Library and the Association
of Canadian Map Libraries is extremely useful for geological libraries
cataloguing maps for an automated cooperative system. With the
inclusion of the Library of Congress MARC Map data base on the OCLC
Map Cataloguing Sub-System from June 1978, Library of Congress
records can be used to check the form of geographic names,
authorities, topical subject headings etc. The most direct and
useful access to Library of Congress MARC Map records, however, must
await OCLC's development of a map classification code search facility.

g. Improvement and standardisation of subject access

As users of a library catalogue geologists and geographers are
often interested in specific subjects on a worldwide basis. The
primary Library of Congress subject pattern Subject - Area accommodates
this interest. However, geologists also employ an areal or regional
oriented methodology concerned with multiple subject aspects of a
specific area or region. Additional subject access under Area-Subject
is needed to accommodate this approach. If both subject access
requirements are to be accommodated within a formalised map
cataloguing programme it follows that both approaches must be
emphasized within the system. A double-entry concept for resolving
the dichotomy between two necessary approaches to area oriented
thematic material is not really new. Double entries under subject
and area were used in general research libraries before 1900 (Frozio,
1973) and are currently being used in a few special area - oriented
libraries (Frozio, 1973).

Although the concept of double entry was, and is, valid for all
place oriented materials, research libraries in the United States,
including the Library of Congress, discontinued or did not adopt it
for economic reasons during a time when unit card sets had to be
typed or set in type for printing. This meant that the standard
Library of Congress subject cataloguing practices as described in
the Library of Congress 'Red Book' have been inadequate and
inconsistent in terms of providing access to area oriented material,
including maps. Until recently all Library of Congress subjects
were treated in one of the following patterns:

1. Subject (Undivided)
2. Subject - Base area - Sub area (Indirect)
3. Subject - Local area (Direct)
4. Subject - Sub area - Subject subdivision
5. Local area - Subject (Applied to local history material and
   recently to maps.)
6. Area - Certain specified subjects.

Although the prevalent Library of Congress Subject Cataloguing
Division pattern for the division of geology oriented subject
headings ('Indirect' Subject Base area - Sub area) is basically
responsive to researcher needs, fragmentation within the total system
and inconsistencies in the application of place to subject headings
mean that neither area nor topical subject are accessible uniformly
or consistently. The Library of Congress treatment of subject

headings for area - oriented material did not meet the information
retrieval requirements of geologists. Recently innovations have been
made such as dividing undivided subject headings 'Indirect' and
converting headings divided 'Direct' to 'Indirect'. These actions
are indicative of the preparation for area - subject rotation or
permutation for implementation at the time Library of Congress
bibliographic access becomes fully automated.

In the interest of uniform access to information, the style,
form, arrangement and content of subject headings assigned to MARC
Map records produced in the Library of Congress Geography and Map
Division have been compatible with the standard Library of Congress
subject heading system. In January 1978 the Geography and Map
Division implemented a system of partial subject heading rotation that
provides additional access under <u>Specific Place-Subject.</u>

The OCLC Map Cataloguing Sub-System provides participating
geological libraries with options and alternatives for improving
the subject control of their collections. Although OCLC currently
does not have on-line subject search capability they are cooperating
with Battelle Institute in the evaluation and testing of a subject
- search capability through the use of minicomputers (Federal Library
Committee, 1977). In serving the more traditional subject catalogues
the OCLC Map Cataloguing Sub-System provides options for the entry
and production of either subject headings completely compatible
with Library of Congress or local subject heading options, either
free text or modified LC subject headings. Cooperating geological
libraries could maximise the reference effectiveness of traditional
subject card catalogues by manipulating and partially permutating
standard Library of Congress subject headings, thereby providing
comprehensive, uniform access to both area and topical subject.

h. <u>Extension of cataloguing coverage to include geological maps in
monographs and serials.</u>

As mentioned earlier geological maps published in monographs
and serials remain an underused, unidentified resource in the
geological library. Locating such maps under the restrictions
caused by present cataloguing practices calls for experienced
reference librarians who have a talent for serendipity, if not
clairvoyance. Such maps should be considered to be integral parts
of the information resources of geological map libraries and should
receive the same level of cataloguing or bibliographic control as
maps published separately. Sharing the costs of cataloguing

separately published maps might enable cooperating libraries to use their cost savings to extend their cataloguing coverage to maps in monographs and serials.

i. Improvement of network capabilities

The most effective approach which cooperating geological map libraries could use for improving the technical capabilities of networks would be to advise and encourage the individual networks and the national bibliographic agencies to accelerate implementation of technical capabilities most responsive to their map cataloguing requirements. Implementation of a geographic-coordinate search capability would be extemely valuable for access to cartographic information contained in map records. This is a proven technique which could be implemented by networks with relatively little software modification. All that stands between the implementation of such a capability is the lack of demand by users. Activation and implementation of MARC Map Format Field 265 for acquisitions information would be useful to geological map libraries. This field is scheduled to contain the name and address of distribution agencies and the listed sale price. Although this is not cataloguing information in the traditional sense, combined with specific area - subject search capability it would be very useful in improving map library acquisitions programmes.

Development of the MARC format for analytical and multi-level capabilities, including activation of the linking numbers concept for typing related records together would improve map libraries' capabilities for cataloguing related texts and maps, sheets of multi-sheet works, maps in monographs and serials etc.

Acceleration of information exchange between data bases on a national or international basis should be encouraged. Exchange or transfer of map records on this basis would prevent duplication of effort on an international scale and would facilitate progress toward increased international standardisation.

REFERENCES

Bibliografie van in Nederland verschenen Kaarten. 1975. The Hague :
Koninklijke Bibliotheck

BRITISH STANDARDS INSTITUTION. 1975. Recommendations for bibliograph-
cal references to maps and charts. Part 1 : References in
accessions lists. BS 5195 London: BSI, 7p.

BUNDESANSTALT FUR BODENFORSCHUNG. 1971. Geologische Ubersichtskarte
1:125 000 Baja Verapaz und Sudteil der Alta Verapaz (Guatemala)
Hannover: Bundesanstalt für Bodenforschung.

CARRINGTON, D & MANGAN, E. 1971. Data preparation manual for the
conversion of map cataloguing records to machine readable form.
Washington : Library of Congress.

DAEHN, R.M. 1975. Maps – the regional approach : a system to share.
Bulletin SLA Geography and Map Division No. 100   p.74.

DUNHAM, K.C. 1967 Practical geology and the natural environment of
Man – 1. Continents and islands. Journal of Geological Society of
London 123 : 1-24.

FEDERAL LIBRARY COMMITTEE. 1977. Newsletter No. 100 : p.1.

ERLACH, A.C. 1961. Geography and map cataloguing and classification
in libraries. Special Libraries 52 : 248-251.

FROZIO, E. 1973. Unpublished memorandum. Washington: Library of
Congress Subject Cataloguing Division.

FULTON, P. & McINTOSH, W.L. 1976. Data Base for Geomap Index.
Proceedings of the American Congress on Surveying and Mapping,
Fall Convention, Seattle, Washington September 28-October 1 pp.329-
337

Geo Katalog   International. 1978. Stuttgart:Geocenter.

Geologic Map of the Arabian Peninsula. 1963. Miscellaneous
Geologic Investigations USGS Map I-270A

GUETTARD, J.E. 1746. Carte minéralogique de la France et
l'Angleterre. Memoires de Mathematique et de Physique de l'Academie
Royal des Sciences. Plate 52.

GUETTARD, J.E. 1752 Carte minéralogique de la Suisse. Memoires de
Mathematique et de Physique de l'Academie Royal des Sciences. Plate 8.

HEEZEN, B.C. & FORNARI, D.J. 1975. Geological map of the Pacific Ocean 1:35 000 000. Initial Reports of the Deep Sea Drilling Project 30.

HILL, J.S. 1977. Developments in map cataloguing at the Library of Congress. Special Libraries 68: 150.

INTERNATIONAL GEOLOGICAL CONGRESS. 1876. First circular Boston. 3p.

INTERNATIONAL GEOLOGICAL CONGRESS. 1881. 2nd Session, Bologne p.104.

INTERNATIONAL GEOLOGICAL CONGRESS. 1913. Commission de la Carte Géologique International de l'Europe. 12th Session Proceedings Ottawa. p.141-143.

JONES, A.G. 1961 Reconnaissance geology of part of West Pakistan. Toronto : Government of Canada. 550p.

LEA, G. 1977. Presentation at the Special Libraries Association Conference New York. June 5-9.

NORTH, F.J. 1928. Geological maps : their history and development, with special reference to Wales. Cardiff : National Museum of Wales vi + 134p.

PARR, T. 1975. Automation of cartobibliography : review of MARC for map library information retrieval and cartographic bibliography. Bulletin SLA Geography and Map Division No. 100 : 38

PITCHER, W.S. & SPENCER, M.O. 1972. Solid geology of north west and central Donegal. In PITCHER, W.S. & BERGER, A.R. The geology of Donegal : a study of granite emplacement and unroofing. London : Wiley Map 1.

RHIND, D. 1977 Cartography : new tasks for the OS. Geographical Magazine 50: 792-793

TANAGLIA, R. 1977. Catalog card filing arrangement for a geology map collection. Bulletin SLA Geography and Map Division No. 110 : 17-23.

UNESCO. 1977 Geological World Atlas. Paris

WALKER, R.D. 1977. Data base review : Georef (plus other geoscience data bases).
Online 1 : 74.

WINCH, K.L. Editor. 1976. International maps and atlases in print. Bowker : London, &c. xvi + 866p.

THE LINK BETWEEN "DATA" AND "DOCUMENTATION"

CORNELIUS F. BURK, Jr.

Canada Centre for Geoscience Data

Department of Energy, Mines and Resources
Ottawa, Canada, K1A 0E4

Summary : "Data" are the smallest indivisible units of
information utilized or produced in the context of specific
intellectual activities; "documentation", on the other hand,
refers to the creation of recorded information (including
"data") and the subsequent control, analysis and dissemination
of knowledge concerning recorded information. Apart from this
semantic link, there are a variety of functional and
organizational commonalities which should and do serve to link
groups that usually deal with "data" and "documentation"
activities in mutual isolation. On the premise that all
information should be managed as a resource in support of
decision-making in substantive areas, these links should be
strengthened through the integration of "literature" and
"data" secondary services, the coordinated management of total
information resources and the sharing of expertise in the
application of information technology.

Concepts and definitions

Much of the terminology associated with information (including
the word "information" itself) has become obscured by multiple
and ambiguous usage. It is necessary at the outset to attempt
some clarification, even at the risk of provoking more
confusion!

With regard to the term "data", there have been various
attempts to define this word in general terms, with the
apparent purpose of isolating, within the entire spectrum of
knowledge, a certain portion which can be labelled objectively
as "data" (e.g. Kotani, 1975; van Olphen, 1975). This seems a
futile approach, since the essence of the concept of "data"
arises from the organization or structure of information,
rather than from any intrinsic quality. "Data" may be defined
as the smallest indivisible units of information utilized or
produced in the context of a specific intellectual activity.
What is or is not "data" depends on the context in which this
information is used or produced. For instance, from the point
of view of those working within the framework of the
scientific method, certain information is accepted or rejected
as "data" depending on established standards of reliability
and reproducibility set by the user and accepted by his or her
peers.

This concept of data is implied in such expressions as:
"Data, as with beauty, are in the eyes of the beholder" (Burk,
1971), or: "One man's information is another man's data".

The idea was well illustrated by Williams' (1975) diagram, reproduced here as Figure 1. Depending on one's level in the hierarchy, whether for example, a surveyor, geologist, exploration manager or legislator, the information may be treated as _either_ information or data. Thus, without knowledge of the specific context and purpose, it is not possible in principle to identify "data".

The term "documentation" refers to the preservation of knowledge. This activity is usually regarded as something within the exclusive domain of librarians and bibliographers or, in an even more restricted sense, associated with computer programming, litigation or other specialities. However, the term has a wider import, and deserves broader recognition as one of the basic activities in the field of information science. "Documentation" may be defined as those activities associated with the creation of recorded information and the subsequent control, analysis and dissemination of knowledge concerning recorded information.

Based on these concepts, the link between "data" and "documentation" can be seen, not as one joining two different _types_ of information as perhaps implied in the title of this paper, but rather as one relating two different aspects of information management, one dealing with a specialized sub-set of information (data) and the other with a sphere of activity (documentation).

Functional links

There are at least four functional areas which link "data" and "documentation" activities:

Bibliographic control of information sources. All documents, if they are worth preserving at all, should be placed under some kind of bibliographic control. However, it must be recognized that "documents" assume a wide variety of forms and contain a broad range of types of information. A magnetic tape or disc file containing scientific observations and measurements should be identified, catalogued and indexed in a manner similar to that traditionally used for conventional books and journals. As more information becomes stored in machine-readable and other novel forms, the need for such expanded bibliographic control becomes greater; happily, several organizations are now involved in the bibliographic control of computer-processable information and other non-conventional media (e.g. Hubaux, 1972; Shapley, 1978). The knowledge and methods of the classical bibliographer will find a ready market in the computer age, if only he or she can focus more on the content and less on the media!

Database administrator and librarian. The age of computers and database management has spawned a new professional, the "database administrator". This individual is responsible for knowing what information is in his or her machine-readable files, where each element of information is, how frequently the files have been updated, and the implications of modifying one part of a file on the remainder. Although the technical functions are different, he or she performs the same job in organizing machine-readable information as does the librarian in organizing and managing a collection of books, journals and maps.

Librarians are of course becoming familiar with the structure and content of various data bases, particularly bibliographies and related secondary files; shelf-spacing and the structural condition of library buildings must be matched with knowledge of file structure, linkages and record lengths.

Information technology. In a world of rapidly advancing information technology, all groups involved in the information field have a common interest in new developments and improvements. Documentalists and data managers alike share interests in areas such as database management, telecommunication networks, text-editing, photocomposition, systems and programming, intelligent copiers, laser platemaking etc. Particular groups tend to make major use of certain types of technology and acquire expertise and familiarity with them which, if available, could be utilized to advantage by other groups. For example, those in documentation are often conversent with the application of text-editors and photocomposition methods, but perhaps not knowledgeable in database management systems; closer liaison and communication are required.

Within an organization the total cost of information technology is usually significant and good management demands that these resources be shared by all groups. Such economic and organizational pressures are forcing what were previously fragmented and disparate information activities to move closer together.

Information management. In recent years there has been increasing recognition given to the concept of "information management" (Horton, 1974; Carlson, 1977); accordingly, information is recognized as a resource and managed in ways analogous to those used for personnel, finances and materiel. Traditionally, most "information" is not managed at all, but simply taken for granted, with the costs buried in overhead! Implicit in the concept of information management is the notion that information resources within an organization are managed in their totality. "Data", "documents" and whatever other information constitutes the resource base must be considered collectively.

Arbitrary distinctions between "data" and "documents" can be misleading and counter-productive. What matters is not the form of the information, but the effectiveness with which these resources contribute to better decision-making.

Organizational links

The natural and logical links between "data" and "documentation" can be illustrated by the current activities of six organizations:

COGEODATA: The Committee on Storage, Automatic Processing and Retrieval of Geological Data (COGEODATA) was established by the International Union of Geological Sciences in 1967 with the main purpose of promoting the use of computer-oriented methods for handling geological data, to promote the standardization of geological data and to develop an index to sources of computer-processable information (Hubaux, 1973).

The objectives and functions of COGEODATA were reviewed during 1977 and will be published as part of a long-range plan for the committee. The newly formulated objectives are:

1. To assess computer-oriented information technology and to promote its worldwide application to the management and interpretation of geological data.

2. To facilitate the collection, compilation and communication of computer-processable geological data.

3. To promote general awareness of data and other information resources in geology, and

4. To provide advice in training assistance within the scope of COGEODATA objectives and activities.

To achieve these goals, the committee has established a number of working and task groups (Burk, 1977).

1. Selection and capture of geological data
2. Data structure and data management
3. Data display
4. Application of remotely sensed data
5. Communication of geological data
   5.1 Retrospective databases in petrology
   5.2 Exchange of geochemical data
6. World index of geological data sources
7. Educational activities

The Working Group on a World Index is concerned with the compilation on an international basis of information on the existence, location and general nature of published data compilations and the major organizational sources of data in the field of geology; as such this work relates most directly to the interests of documentalists (Hubaux, 1972).

CODATA, International Council of Scientific Unions: Whereas COGEODATA serves the geological sciences, CODATA (Committee on Data for Science and Technology) is an inter-disciplinary committee (involving geologists) of the International Council of Scientific Unions. It was formed in 1966 to deal with data of importance to science and technology, their compilation, critical evaluation, storage and retrieval; its scope now includes quantitative data on the properties and behaviour of matter, characteristic of biological and geological systems, and other experimental and observational data (Westrum, 1977).

CODATA's purpose is to promote data compilation and evaluation, to improve the quality of data collections and their usefulness to the user community, and to improve data accessibility.

Two main CODATA activities are of particular relevance to documentation. The Task Group on Accessibility and Dissemination of Data (ADD) is concerned with reducing the barriers between data generation and possible application and use. Such barriers include unawareness, language and vocabulary confusion, failure of communication channels, differing needs and different disciplines, geographical separation, currency problems, lack of coordination and so on. The task group has acquired an appreciation of the broad range of scientific and technical data. It has learned that different disciplines acquire, store, evaluate and apply data in different ways; and discovered that many separate organizations are involved in data accessibility and dissemination, often with no clear definition of responsibility among those taking part.

The World Data Referral Centre (WDRC) was established with support from UNESCO at CODATA Headquarters in Paris in 1977, as a result of work by the ADD Task Group (UNISIST Newsletter 5/2:6). Functions and services of WDRC include collecting information concerning major recources of data on a worldwide basis; preparing and disseminating documents to assist local and national services to conduct data referrals; performing data referrals on request from national and local services and, where necessary, from individual inquirers. Related to this function is the compilation of CODATA's "Directory of Data Sources for Science and Technology".

The CODATA directory has been planned as a successor to the international compendium of numerical data projects published in 1969 as one of CODATA's (1969) first major tasks. The new directory will be divided into a number of chapters, each representing a single scientific discipline or group of related disciplines. Individual chapters will be published as an issue of the CODATA Bulletin (e.g. Watson, 1977). Responsibility for compilation of the chapter on geology was accepted by COGEODATA and the Working Group on a World Index, previously described, will contribute its results.

In the context of this directory, the term "data" is taken to mean "quantitative data" on the properties and behaviour of matter, quantitative data and characteristic values of biological, geological and astronomical systems, and other experimental and observational values. Information sources and data bases that are strictly bibliographical in character will not be included.

Food and Agriculture Organization of the United Nations (FAO): The Remote Sensing unit of FAO, Rome, Italy has undertaken to investigate the feasibility of developing a World Index of Space Imagery (WISI). This project is another example of the need to combine the skills of the data manager and the documentalist. The growing number of earth-orbiting and stationary satellites are producing digital data on planet Earth in truly astronomical quantities! Using the most sophisticated data storage technology available, even large warehouses are becoming insufficient for storage. The implications of such problems, associated with the rate at which data are now being generated, are being viewed seriously by some in terms of imposing a limit to growth (Hartmann, 1978).

From a documentalist's point of view, one of the principal problems posed by WISI is the need for a world grid for indexing purposes. Although certain individual projects and organizations utilize various global grids for indexing purposes (not to be confused with the direct utilization of latitude and longitude coordinates), there is no generally acknowledged, well-known grid which could be applied by indexers on a standardized global basis for geological, environmental and other space-dependent data. This is a problem common for both data managers and documentalists and one which could be resolved through cooperation by all countries on a multi-disciplinary basis. Perhaps geologists should take the lead.

Geosystems: Another logical and operational link between "data" and "documentation" has been established by Geosystems, through the extraction of certain categories of data from documents referenced in the Geoarchive bibliographic file (Lea, 1978). Examples include the extraction of data from geological maps and mineral production data from published papers. The map data example is based on Geosystems' work with 100 000 maps held by the (U.K.) Institute of Geological Sciences.

ENDEX: The Environmental Data Service of the U.S. National Oceonographic and Atmospheric Administration includes among its services a file called ENDEX (Environmental Data Index), designed to provide rapid, computerized referral to available environmental data files. The databases can be searched by geographic area, measured parameters, institution holding the data, and so on (Hughes, 1976).

Canadian Index to Geoscience Data: The first major effort in geology to bridge the gap between "documentation" and "data" is represented by the Canadian Index to Geoscience Data. Begun in 1967, this project has had as its primary objective the bibliographic control of documents containing observations and measurements (data) dealing with the Canadian landmass and its offshore regions. The Index was conceived as a supporting element for a proposed national system for the storage and retrieval of primary geological data, and was intended to complement the general literature-based bibliographic services such as GeoRef and Geoarchive (Gunn and Burk, 1975; Burk, 1978).

The thesaurus of the Canadian Index, used for vocabulary control of indexing terms, can be regarded as a list of "data tags" (Lloyd, 1976). Particular attention is paid to geographic control, and for these purposes a national grid called the National Topographic System (NTS) is utilized for indexing all documents. Experience by users indicates that the Canadian NTS geographic grid is the single most useful searching parameter. An international grid would have been preferred, but none was readily available.

Conclusions

By way of summarizing, there are three main areas in which the link between "data" and "documentation" can be strengthened:

1.   Integration of "literature" and "data" secondary services

Existing services could widen their scope to include both types;

since "data" collections and files tend to be unpublished and not duplicated, a new network mechanism is required to identify and update secondary information dealing with these sources;

"data" collections should be more extensively advertised, as for example through the CODATA Directory; and

regional and global systematic grid systems should be used as common keys for linking bibliographic and data sources.

2.   Management of total information resources

Recorded information should be recognized as an identifiable, manageable resource and the importance of information management should receive greater attention;

all information activities - whether "data" or "document"-oriented - should be managed to help achieve national, organizational or other goals; information activities have value only to the extent that they conserve other resources through better decision-making; and

through re-organization, information-related activities should be brought closer together; for example, the IUGS Committees on Data (COGEODATA) and Geological Documentation should be unified.

3.   Sharing information technology

Most information technology, computers, telecommunication networks, word processors, and so on, is applicable to all forms of information, and expertise on the application of such technology can and should be shared by both "data" and "document"-oriented groups in the geological sciences; and

the cost and management of this resource should be shared in the interests of better utilization of funds and expertise.

References

BURK, C.F., Jr. 1971. Computer-based geological data systems : an emerging basis for international communication: Proc. Eighth World Petroleum Congress 2 : 327-335.

BURK, C.F., Jr. 1977. Editor. Progress reports for COGEODATA working groups, 1977. COGEODATA Newsletter 3/4 : 1-8.

BURK, C.F., Jr. 1978. The national data referral system for Canadian geoscience. Geoscience Information Society Proceedings 8 : 31-41.

CARLSON, W.M. 1977. Where is the pay off? American Society for Information Science Bulletin 4 : 14-17.

CODATA. 1969. International compendium of numerical data projects : New York : Springer-Verlag, 295pp.

GUNN, K.L. and BURK, C.F., Jr. 1975. The Canadian Index to Geoscience Data : a decentralized, cooperative indexing project. Canadian Association for Information Science Proceedings 3d Conference, Quebec, 242-253.

HARTMANN, G.K. 1978, The information explosion and its consequenses for data acquisition, documentation and processing : an additional aspect of the limits to growth : World Data Centre A for Solid Earth Geophysics, U.S. Dept. Commerce, NOAA, Rept. SE-11, 36pp.

HORTON, F.W. Jr. 1974. How to harness information resources : a systems approach. Cleveland : Association for Systems Management, 147pp.

HUBAUX, A. 1972. Geological data files - survey of international activity. CODATA Bulletin 8 : 30pp.

HUBAUX, A. 1973. Report for the first term 1967-1972 [of COGEODATA]. Geological Newsletter 1973 : 130-134.

HUGHES, P. 1976. Editor. New ENDEX services : EDS (Environmental Data Service) March 1976 : 24.

KOTANI, M. Chairman. 1975. Study on the problems of accessibility and dissemination of data for science and technology. CODATA Bulletin 16, 31pp.

LEA, G. 1978. Geoarchive : Geosystems indexing policy. Geoscience Information Society Proceedings 8 : 42-56.

LLOYD, J. Chairman. 1976. Flagging and tagging data. CODATA Bulletin 19, 22pp.

SHAPLEY, A.H. Chairman. 1978. Directory of U.S. data repositories supporting the international geodynamics project. World Data Centre A for Solid Earth Geophysics, U.S. Dept. Commerce, NOAA, Rept. SE-14, 40pp.

VAN OLPHEN, H. 1975. What are "data"? American Society for Information Science Bulletin 1/7 : 8-9.

WATSON, D.G. Editor. 1977. CODATA directory of data sources for science and technology. Chapter 1, Crystallography. CODATA Bulletin 24, 42pp.

WESTRUM, E.F., Jr. 1977. Editor. CODATA. Paris : CODATA Secretariat, 44pp.

WILLIAMS, G.D. 1975. Advantages of using a generalized system
to manage geological data. In HUTCHISON, W.W. Chairman.
Computer-based systems for geological field data.
Geological Survey of Canada Paper 74-63 : 84-88.

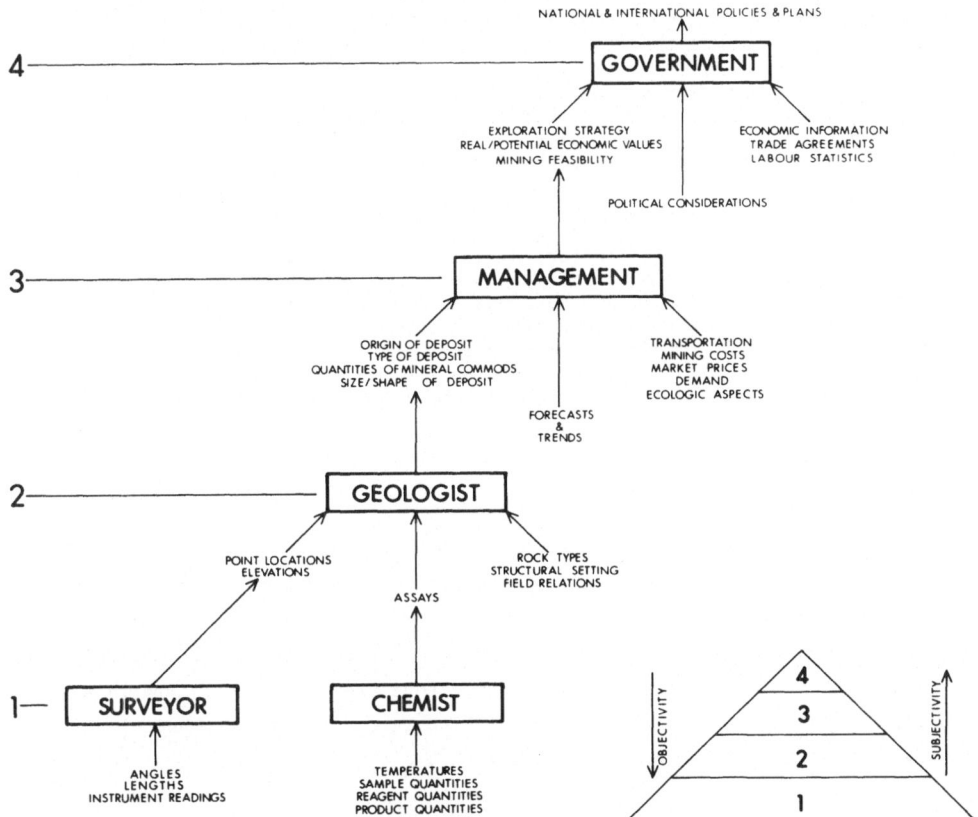

Figure 1. Hierarchy of "data" and "information". **Reproduced from Williams (1975 : 85)**

EARTH SCIENCE AND THE FOREIGN LANGUAGE PROBLEM

D N WOOD BSc, PhD
Head of Acquisitions
British Library Lending Division
Boston Spa, Wetherby LS23 7BQ

Summary: The language problem facing earth scientists is probably greater than in other scientific disciplines. Geology and related subjects are studied and researched throughout the world and workers in the field probably have a more pressing need than others to keep abreast of developments outside their own geographical region and language area.

Recently conducted surveys at the British Library Lending Division show that compared with other subject fields only a relatively small proportion of earth science papers are written in English. Furthermore, the linguistic expertise of earth scientists, at least in the UK, hardly fits them to cope with the mass of foreign literature of potential value. Russian and German literature causes particular problems.

To overcome the language barrier English speaking earth scientists receive more help than most. Hundreds of individual articles are translated each year and these are recorded in card indexes and published bibliographies. Cover-to-cover translations, in many cases going back several years, are helpful in keeping up-to-date particularly with Soviet literature. Translation agencies are on-hand to help (at a price) in the event of a translation not being available.

The Problem

The literature of the earth sciences is probably as scattered as that of any other discipline. Geology, like agriculture and medicine tends to be studied throughout the world and with the growing tendency for books and journal articles to be published in the vernacular there is an increasing number of languages represented in the literature with which librarians, information officers, practicing geologists, research workers and, to a lesser extent, students have to deal.

Having regard to the numbers of papers indexed in the 1977 issues of Bibliography and Index of Geology, the earth science sections of Bulletin Signalétique, and the different sections of Referativnyi Zhurnal which are relevant to the study of geology and related subjects, the total number of documents published each year of potential interest to earth scientists is well in excess of 50,000 and is probably nearer 80,000. An analysis of a sample of references in these three guides reveals the following language breakdown: English 55.5%, Russian 23.5%, French 8.5%,

German 5%, Japanese 1% and the remaining languages 6.3%. It should be
mentioned that these figures are at variance with those contained in a
paper by Hawkes in 1967. He analysed a much larger number of secondary
services and concluded that only 27% of around 30,000 articles published
in 1961 were in English, 30% in Russian, 11% in French, 11% in German,
9% in non-Russian Slavic languages, 11% in other European languages and
1% in non-European languages. Accepting these lower percentages however,
still means that there are at least 28,000 earth-science articles per
year published in languages other than English.

Despite the large amount of foreign language material published citation
analyses indicate that no great use is made of it. In an analysis of the
references in a number of leading British and US geological journals it
was revealed that 89% of the citations by North American authors and 84%
of those by British authors were to English language publications
(Wood, 1973). German at 3.4% and 8.6% respectively was the next most
frequently cited language.

The low level of use of non-English language material can be attributed
to three main causes. In the first place much geological information is
of purely regional interest - thus American authors do not even cite
British publications very much and vice versa; secondly, many earth
scientists are unfamiliar with the guides to the literature and hence
fail to discover the existence of foreign language articles and books
which might be of interest to them; and thirdly, there is the language
problem.

A recently conducted survey of the foreign language problem facing
British scientists and technologists has just been completed (Ellen,
in press). It confirms the findings of a previous survey (Wood, 1967)
and shows that of the earth scientists that replied 63% claimed to be
either fluent or well able to cope with French, 11% with German, 5% with
Spanish and 2% with Russian. Eleven other languages were also mentioned.

The picture presented by these figures is not very encouraging but does
this matter? After all if no-one wishes to read foreign language papers
the fact that they cannot do so is immaterial. To obtain information on
this question respondents were asked whether they had recently come
across a paper which they would like to have read but could not because
of the language problem; and if so, how recently and in what language.

The replies indicate that within the month prior to receiving the questionnaire no less than 45% had encountered a language problem. In a third of the cases this had involved German, in a further third Russian, in 12% of the cases French, 7% Japanese and 6% Polish. Eleven other languages were also mentioned.

From this it can be concluded that, in the UK at least, there is evidence of a considerable language problem among the earth-science community.

Translations Indexes

What can be done to solve this problem? Long term solutions may include firstly the teaching, or in some cases the better teaching, of more languages at school level; secondly the development of special language courses for scientists, for example the intensive course in Japanese which was developed recently at Sheffield University; thirdly the provision within research establishments and universities of services which could enable would-be readers to pick the bones out of an article without going to the trouble and expense of obtaining full length translations, ie a service similar to that being operated by the British Library, Science Reference Library; and finally, the use of computers to produce 'instant' translations.

Although such developments are taking place they are doing so only slowly and it will be some time before most language problems can be solved without the aid of a published, or semi-published, manually produced translation.

Around 20,000 individual articles per year are translated into English and of these about 6% are in the earth-sciences field. In addition, several hundred journals are translated regularly from cover-to-cover. So that these translations may be made available to those who require them several organisations have, within the last 20 years or so, been concerned with either collecting or recording them: some have been doing both.

In the UK the main body is the British Library. At the Lending Division (BLLD) special attention has been paid to developing a collection of translations - this policy has its origin in the National Lending Library for Science and Technology where Russian literature (and translations of it) was one of the main priorities of the new library in the late 1950's.

The collection now includes nearly 450,000 individual article translations from all languages into English, complete files of over 250 journals translated regularly into English, and several thousand translated books. The books and journals are obtained through commercial channels but individual articles are usually acquired direct from the originators, for example Government departments and agencies, industrial firms, universities and research organisations. They include about 3,000 per year from British sources and about 15,000 per year from American organisations, including the National Translations Center, National Technical Information Service, and the Joint Publications Research Service. All translations received at the Lending Division are indexed in card catalogues. Article translations are indexed under original journal title and books under author. These indexes enable the Division to provide an information service on the availability of translations and the Library is currently dealing with around 40,000 enquiries annually. The success rate is about 20%.

In addition to collecting and indexing translations the Lending Division publicises its holdings through its monthly publication BLLD Announcement Bulletin, although it only contains details of British translations. This will eventually be discontinued since it is expected that UK translations plus other material such as reports and dissertations will soon be the subject of a supplement to the British National Bibliography.

The only other major multi-disciplinary collection of translations in the UK is maintained at the British Library, Science Reference Library. Over 30,000 individual article translations are held and the Library subscribes to most of the commercially available journal translations. A card index is available and where translations are held, the original is also marked to alert readers to the fact.

Another organisation in the UK concerned with translations is Aslib. It does not collect translations but maintains a card index giving details of where they can be obtained. This index started life as an international cooperative project known as the Commonwealth Index of Unpublished Translations; the Commonwealth aspect however faded out in the early to mid-70's. It contains details of around 500,000 translations of patents, standards, journal articles, books, reports and theses but can be used only by Aslib's member organisations.

In the US, most of the multi-disciplinary translation activity centres around two organisations - the National Translations Center and the National Technical Information Service (NTIS). The former is based at the John Crerar Library in Chicago where a collection of translations is maintained and a monthly index published. This index - the Translations Register-Index - gives details of all the translations acquired by the Center as well as additional translations made available through NTIS. Translations Register-Index cumulates quarterly and annually. A useful tool to use in conjunction with Translations Register-Index is the Consolidated Index of Translations into English, a bibliography of around 160,000 English translations announced during the period 1954 to 1966 ie up to the date when Translations Register-Index commenced.

In view of the similar terms of reference of the British Library Lending Division and the National Translations Center, ie to collect translations from all languages into English, a close working relationship has been established between the two organisations. An exchange of translations and index cards take place and the bibliographical tools produced by NTC function effectively as guides to the Lending Division's holdings.

The principal producers of translations in the Western world are the United States Government and its various agencies. Their output is made available through the National Technical Information Service and in addition to being listed in Translations Register-Index the translations are announced in Government Reports Announcements and Index, the twice monthly bibliography of American research reports available through NTIS. A subset of this material - the 30,000 translations issued per year by the Joint Publications Research Service are also announced separately in a publication known as Transdex Index. These translations include many earth science items, among them collections of abstracts of Russian and Chinese material.

National translation centres have been established in many other countries, their primary function usually being to collect and index translations into their own languages. In France it is the Centre National de la Recherche Scientifique (CNRS) which has this responsibility; in West Germany the Technical Library of the Technical University in Hanover; and in East Germany the German Academy of Sciences in Berlin.

Some European countries, and indeed some countries outside Europe, support the operation of the International Translations Centre (ITC) which is housed in the University of Delft in the Netherlands. Its predecessor, the European Translations Centre (ETC), was established in 1960. ITC is charged with collecting, indexing and publicising translations from the difficult languages (Russian, Japanese, Chinese, Rumanian, Bulgarian, Czech, Polish, Serbo-Croat etc) into any of the languages of Western Europe. In the near future however, it seems likely that inter-Western translations will also be covered. Since 1967 ETC/ITC has published a World Index of Scientific Translations. This is now issued monthly with regular cumulations and this year has been combined with the monthly index of the CNRS and the more specialised Transatom Bulletin (published by Euratom and covering atomic energy) under the title World Transindex.

All the centres mentioned so far will deal with enquiries, and can supply information about, or copies of, translations recorded in their indexes. English-speaking workers are advised to make use of BLLD or NTC depending upon whether they reside in the UK or the USA. For English language material ITC can offer no better service but may be useful to contact where translations into German or French, for instance, would be acceptable.

Many of the translations appearing in these lists and card indexes are generated as part of translations series produced by companies, learned societies and government departments. Many of them are subject oriented. Within the earth science field there are, for instance, the translations series of the American Geological Institute, the American Geophysical Union, the Arctic Construction and Frost Laboratories, the Institute of Oceanographic Sciences, the Meteorological Office, the National Oceanic and Atmospheric Administration, the Smithsonian Institution, the Snow, Ice, Permafrost Research Establishment and the US Bureau of Mines. Many of these organisations produce lists of their translations but these are useful mainly as current awareness tools since the material they contain is eventually incorporated into the main translation indexes.

A point to note about most of the translations available from or through the channels referred to above is that they are in semi-published form. Many are available as xerox or offset copies made from typed manuscripts of variable quality. The majority, however, are distributed in the form of 35 mm microfilm or, in the case of the more recent items, as

microfiche. Enlarged copies of these filmed versions can usually be
provided by the libraries or translations centres which hold them but
the cost can be quite considerable in the case of a long translation.

Cover-to-Cover and Selective Translations

Brief mention has already been made of the existence of cover-to-cover
translations. These are regularly produced translations of complete
issues of certain foreign language journals. There are currently over
250 scientific and technical periodicals being translated. Most are
translations from Russian but a few Polish, Japanese and Chinese and at
least one German journal are also translated from cover-to-cover. They
constitute a valuable means of keeping abreast of overseas developments.
Unfortunately, because of inevitable delays in acquiring the original
journal, translating it, and editing and publishing the translation,
most cover-to-cover translations appear several months and sometimes
years after the original.

In recent years there has been a growing realisation that much of the
material appearing in cover-to-cover translations is of little value to
readers and that, in certain fields at least, a more discriminating
programme of translations was called for. As a result there has been
a gradual increase in the number of selective translation journals.
These are of two types, those where only a proportion of the articles
in the original appear in the translation and those which bring together
collections from several sources.

A list of cover-to-cover and selective translation journals in the earth
sciences field is given in Appendix 1. Further details of these and
other translated journals appear in several bibliographies. The latest
to appear is Translations Journals, a joint venture of the British
Library Lending Division and the International Translations Center; it
was published in 1978.

In the case of Japanese material much selective translation work is
undertaken in Japan itself; presumably with the object of promoting
Japanese science and technology in Western countries. Unfortunately a
large proportion of the translations are not subject to adequate bibliographical control and many articles are published in English without any
reference to the fact that they are translations of Japanese articles

from other journals. A list of the English language earth sciences journals published in Japan is presented as Appendix 2.

Translations journals and the various bibliographical tools mentioned above are unfortunately not well used. The results of the survey mentioned earlier show that despite these tools, when faced with a language problem less than 10% of earth scientists try to locate a translation - probably because few are aware of the guides that do exist. For instance only 14.8% had heard of World Index of Scientific Translations and only 5.6% of Translations Register Index. Librarians and information officers should ask themselves whether they publicise such bibliographic tools frequently enough.

Book Translations

Many thousands of scientific and technical books are published each year throughout the world. Most of these are, like English language books, in the nature of review publications and many have got English language equivalents, a fact which obviates the need for translations. However, this is perhaps less true in the earth sciences - particularly in the case of regional geology for instance - than in other subject fields and this makes book translations important sources of information.

Most book translations are produced commercially and are announced in the national bibliographies of the countries in which they are published but there are also a number of special guides. Several of those mentioned previously index translated books to some extent but the most comprehensive list is Index Translationum. This has been published annually since 1949 by Unesco although it is rather belated in its appearance; the latest issue covers translations published in 1974. A cumulated index to those translations published in Australia, Canada, the United States of America, Ireland, New Zealand, South Africa and the United Kingdom and listed in Index Translationum between 1948 and 1968 has been published by G K Hall under the title Cumulative Index to English Translations. The work is in two volumes and lists some 44,000 translations in a single alphabetical sequence according to the name of the author.

Incidentally the British Library Lending Division attempts to acquire all worthwhile books translated into English and currently possesses around 20,000.

Getting a Translation Prepared

Despite the existence of many thousands of article and book translations the bulk of the world's literature remains only in its original language. Even several years after its publication the chances of a translation of any particular item being available are only about 1 in 5. If therefore one discovers the existence of a foreign language paper which is suspected of containing valuable information, and a translation cannot be located, what is the next step?

If only the title is available it may be worth obtaining a copy of the original paper. Many journals publish English abstracts of the papers they carry and some, particularly Japanese journals, contain translated summaries and/or conclusions and carry English captions to the figures and tables. In most cases it will then become obvious whether the paper is really going to be useful. It may be that the tables, figures and conclusions are all that are required anyway.

Should the article itself not offer any help to the English speaking reader it might be worthwhile searching for an abstract in an English language abstracting journal. Alternatively, linguistic help may be obtainable from a colleague, an information department or, for those in the London area, from the Science Reference Library. If preliminary enquiries suggest that a full translation is worthwhile, translation services may be available within one's organisation (cheap) through a commercial translation burea (expensive), through a free lance translator (cheaper but may be poor) or the British Library Lending Division. The Lending Division produces translations of articles and books on request at subsidised rates. The fee charged represents 50% of the payment to the translator. The remainder is hopefully recovered from subsequent sales. In return for this cheap service users (and at the moment the service is limited to British users) are expected to edit the translation from a technical point of view. The translations so produced are listed regularly in the BLLD Announcement Bulletin.

References:

ELLEN, S. In press. Survey of foreign language problems facing the research worker. _Interlending Review_.

HAWKES, H.E. 1967. Geology. In DOWNS, R.3. and JENKINS, F.B. Editors. Bibliography, current state and future trends. Part 2. _Library Trends_ 15, (4) 816-828.

WOOD, D.N. 1967. The foreign language problem facing scientists and technologists in the United Kingdom - report of a recent survey. _Journal of Documentation_ 23, (3), 117-130.

WOOD, D.N. 1973. Translations and foreign literature. In WOOD, D.N. Editor _Use of earth sciences literature_. London: Butterworths: 104-121.

JOURNALS CONSISTING OF TRANSLATIONS OF ARTICLES SELECTED FROM VARIOUS FOREIGN JOURNALS

Title of Journal	Originating date	Source
Geochemistry International (selected articles from Geokhimiya and other journals)	1964-	AGI
International Geology Review (selected articles from Russian journals)	1959-	AGI
Petroleum Geology	1974-	Petroleum Geology
Polar Geography: Review and Translation	1977-	Scripta
Soviet-Bloc Research in Geophysics, Astronomy and Space (selected articles from Russian and Chinese journals)	1961-1975	NTIS
Soviet Geography: Review and Translation	1960-	AGS
Soviet Hydrology: Review and Translation	1962-	AGU
Soviet Soil Science	1969-	SSSA

COVER-TO-COVER TRANSLATIONS IN THE EARTH SCIENCES FIELD

Title of original journal	Years available in translation	Title of Translation	Source
Antarktika - Doklady Komissii	1960-	Antarctica - Commission Reports	NTIS
Byulleten Moskovskogo Obshchestvo Ispytatelei Prirody, Otdel. Geologicheskii	1956-58	Bulletin of the Moscow Society for Natural Research, Geological Section	NTIS
Dili/Ti-Li	1961-1965	Translations from Ti-Li (Geography)	NTIS
Dili Xuebao/Ti-Li Hsueh Pao/Acta Geographica Sinica	1966 (1)	Acta Geographica Sinica	NTIS
Diqui Huaxue/Ti Ch'iu Hua Hsueh/Geochimica	1973 (1)	Geochimica	Plenum
Diqui Wuli Xuebao/Ti Ch'iu Wuli Hsueh Pao/ Acta Geophysica Sinica	1973 16 (1)	Acta Geophysica Sinica	Plenum
Dizhi Kexue/Ti Chih K'o Hsueh/Scientia Geologica Sinica	1973 1 (1)	Scientia Geologica Sinica	Plenum
Dizhi Xuebao/Ti Chih Hsueh Pao/Acta Geologica Sinica	1973 (1)-(2)	Acta Geologica Sinica	Plenum
Doklady AN SSSR Geokhimiya Geologiya, Geofizika, Petrografiya, Mineralogiya, Gidrogeologiya, Paleontologiya	1959-1961	Doklady of the Academy of Sciences of the USSR: Earth Sciences Sections/ Doklady Earth Sciences Sections	Kraus
Doklady AN SSSR Geokhimiya, Geologiya, Geofizika, Petrografiya, Mineralogiya, Gidrogeologiya, Okeanologiya Paleontologiya	1962-	Doklady Earth Sciences Sections	AGI
Doklady AN SSSR Geokhimiya	1956-1958	Proceedings of the Academy of Sciences of the USSR, Geochemistry Section	Plenum

Title of original journal	Years available in translation	Title of Translation	Source
Doklady AN SSSR Geologiya	1957-1958	Proceedings of the Academy of Sciences of the USSR, Geological Sciences Section	Plenum
Doklady AN SSSR Geologiya	1959-	Doklady Earth Sciences	AGI
Doklady AN SSSR Okeanologiya	1961	Doklady of the Academy of Sciences of the USSR, Oceanology	Scripta
	1962-1964	Soviet Oceanography	Scripta
	1965-	Doklady Earth Sciences Sections	AGI
Doklady AN SSSR Pochvovedenie	1964-1968	Soviet Soil Science	SSSA
Fizikotekhnicheskie Problemy Razrabotki Poleznykh Iskopaemykh	1965-	Soviet Mining Science	Plenum
Geodeziya i Kartografiya	1958-1961	Geodesy and Cartography	AGU
	1973-	Geodesy, Mapping and Photogrammetry (Selective)	AGU
Geofizicheskii Byulleten	1966-	Geophysical Bulletin	
Geokhimiya	1956-1963	Geochemistry (after 1963, selected articles only published in Geochemistry International)	STI
Geologiya i Geofizika	1965- ?	Geology and Geophysics	Allerton
	1974-	Soviet Geology and Geophysics	
Geologiya Nefti i Gaza	1958	Petroleum Geology	Petroleum Geology
	1959-1966	Petroleum Geology (selective)	"
	1973	Petroleum Geology (selective)	"

Title of original journal	Years available in translation	Title of Translation	Source
Geologiya Rudnykh Mestorozhdenii	1960(1)-1962(2)	Economic Geology, USSR (selective)	Pergamon
Geoloski Glasnik	1956 1 -	Geological Bulletin	NTIS
Geomagnetizm i Aeronomiya	1961 1 (1) -	Geomagnetism and Aeronomy	AGU
Geomorfologiya	1970(1)-(4)	Geomorphology	Plenum
Geotektonika	1967 2 (1) -	Geotectonics	AGU
Informatsionnyi Byulleten Sovetskoi Antarkticheskoi	1958-1961 30 1961 31-	Information Bulletin of the Soviet Antarctic Expedition Soviet Antarctic Expedition, Information Bulletin	AEP AGU
Itogi Nauki-Fizika Zemli	1974-	Summaries of Scientific Progress - Physics of the Earth	Hall
Itogi Nauki - Mestorozhdeniya Goryuchikh Poleznykh Iskopaemykh	1973-	Summaries of Scientific Progress - Deposits of Fossil Fuels	Hall
Itogi Nauki - Meteorologiya i Klimatologiya	1971-	Summaries of Scientific Progress - Meteorology and Climatology	Hall
Itogi Nauki - Okeanologiya	1971-	Summaries of Scientific Progress - Oceanology	Hall
Itogi Nauki - Razrabotka Neftyanykh i Gazovykh Mestorozhdenii	1973-	Summaries of Scientific Progress - Development of Oil and Gas Deposits	Hall

Title of original journal	Years available in translation	Title of Translation	Source
Itogi Nauki - Teoreticheskie Voprosy Fizicheskoi i Ekonomicheskoi Geografii	1973-	Summaries of Scientific Progress - Theoretical Problems in Physical and Economic Geography	Hall
Izvestiya AN SSSR, Fizika Atmosfery i Okeana	1965-	Izvestiya, Academy of Sciences, USSR. Atmospheric and Oceanic Physics	AGU
Izvestiya AN SSSR, Fizika Zemli	1965	Izvestiya, Academy of Sciences, USSR: Physics of the Solid Earth	AGU
Izvestiya AN SSSR, Seriya Geofizicheskaya	1957	Bulletin/Izvestiya of the Academy of Sciences, USSR, Geophysics Series	AGU
Izvestiya AN SSSR, Seriya Geologicheskaya	1958-1961	Bulletin/Izvestiya of the Academy of Sciences, USSR, Geologic Series	AGI
	1962-	Translated cover-to-cover, but published selective in International Geology Review	AGI
Izvestiya VUZ Geodeziya i Aerofotosemka	1962-1972 (6)	Geodesy and Aerophotography	AGU
	1973-	Geodesy, Mapping and Photogrammetry (selective)	AGU
Kristallografiya	1956-	Soviet Physics. Crystallography	AIP
Litologiya i Poleznye Iskopaemye	1966-	Lithology and Mineral Resources	Plenum
	1964	Economic Geology, USSR (selective)	Pergamon
Meteoritika	1963 <u>23</u>	Meteoritica	Taurus Press

Title of original journal	Years available in translation	Title of Translation	Source
Meteorologiya i Gidrologiya	1966–1975 (10)	Meteorology and Hydrology (selective)	NTIS
	1976–	Soviet Meteorology and Hydrology	Allerton
Novye Knigi SSSR	1964	Forthcoming Russian Books: Technical Sciences	SIC
	1965–1969	Index to Forthcoming Russian Books: Technical Sciences	SIC
	1970–	Index to Forthcoming Russian Books	SIC
Okeanologiya	1965–	Oceanology	AGU
Osnovaniya Fundamenty i Mekhanika Gruntov	1964–	Soil Mechanics and Foundation Engineering	Plenum
Paleontologicheskii Zhurnal	1962–1966	Translated cover-to-cover but published selectively in International Geology Review	AGI
	1967–	Paleontological Journal	AGI
Problemy Arktiki i Antarktiki	1962(11)–1970(36)	Problems of the Arctic and Antarctic	NTIS
Pochvovedemie	1958–1961	Soviet Soil Science	NTIS
	1962–1968	Soviet Soil Science	SSSA
Prikladnaya Geofizika	1966–1968	Exploration Geophysics	Plenum
Problemy Severa	1958–	Problems of the North	NRCC
Rudarsko-Metallurski Zbornik	1962–1972	Mining and Metallurgy Quarterly	NTIS

Title of original journal	Years available in translation	Title of Translation	Source
Shui Wen Yue Kan/Shui-wen Yuch-K'an	1959(5)(11)(12)	Hydrological Monthly	NTIS
Sovetskaya Geologiya	1960-	Translated cover-to-cover but published selectively in International Geology Review	AGI
Trudy Geofizicheskogo Instituta, AN SSSR	1957 37-40	Soviet Research in Geophysics	Plenum
Trudy Morskogo Gidrofizicheskogo Instituta, AN SSSR	1959 16-17	Soviet Oceanography/Transactions of the Marine Hydrophysical Institute (selective)	Scripta
	1959 18-1964	Soviet Oceanography/Transactions of the Marine Hydrophysical Institute	Scripta
Vesnik Zavod za Geoloska i Geofizicko Istrazivanje	1962-	Bulletin of the Institute for Geological and Geophysical Research - issued in three series	NTIS
Vestnik Moskovskogo Universiteta, Seriya 4: Geologiya	1974-	Moscow University Geology Bulletin	Allerton
Vestnik Moskovskogo Universiteta, Seriya 6: Biologiya, Pochvovedenie	1974-	Moscow University Soil Science Bulletin	Allerton

JAPANESE EARTH SCIENCE JOURNALS PUBLISHED IN ENGLISH

Title of journal	Years available
Annual Report of Hydrological Research in the Area of Lake Biwa	1967 -
Applied Geography (Annual Report, Association of Applied Geographers)	1960 - 1964?
Bulletin of the Department of Geography, University of Tokyo	1969 -
Bulletin of the Earthquake Research Institute, University of Tokyo	1926 - 1971 49
Bulletin of the Geographical Survey Institute	1948 -
Bulletin of the International Institute of Seismology and Earthquake Engineering	1964 -
Bulletin of the Ocean Research Institute, University of Tokyo	1967 -
Bulletin of the Niigata Airglow Observatory	1972 -
Bulletin of the Urakawa Seismological Observatory	?
Bulletin of Volcanic Eruptions	1961 -
Climatological Notes of the Department of Geography, Hosei University	1969 - 1973 (15)
Climatological Notes of the Institute of Geoscience, University of Tsukuba	1974 (16) -
Contributions of the Geophysical Institute, Kyoto University	1971 (11) -
Contributions of the Marine Research Laboratory, Hydrographic Office of Japan	1959 -
Geochemical Journal	1966 -
Geology and Mineral Resources of Japan	1956 -
Geographical Reports, Tokyo Metropolitan University	1966 -
Geology and Palaeontology of S.E. Asia	1964 -
Geophysical Magazine, Tokyo	1926 -
IISEE Lecture Notes, International Institute of Seismology and Earthquake Engineering	1965 -

Title of journal	Years available
Individual Studies by Participants, International Institute of Seismology and Earthquake Engineering	1964 –
Japanese Journal of Geology and Geography	1922 – 1975 45 (1-4)
Japanese Journal of Geophysics	1959 – 1970 5 (1)
Japanese Progress in Climatology	1964 –
JARE Scientific Reports, Series A: Aeronomy	1963 – 1973 (11)
JARE Scientific Reports, Series B: Meteorology	1969 (1)
JARE Scientific Reports, Series C: Geology, Geography, Glaciology, Seismology, Geodesy and Geochemistry	1964 – 1973 (7)
JARE Scientific Reports, Series D: Oceanography	1964 – ?
Journal of Earth Sciences, Nagoya University	1953 –
Journal of the Faculty of Science – Hokkaido University, Series 4: Geology and Minerology	1930 –
Journal of the Faculty of Science, Hokkaido University, Series 7: Geophysics	1957 –
Journal of the Faculty of Science, Niigata University, Series 2: Biology, Geology and Minerology	1952 – 1967
Journal of the Faculty of Science, University of Tokyo, Section 2: Geology, Minerology, Geography, Geophysics	1925 –
Journal of Geomagnetism and Geoelectricity, Kyoto	1949 –
Journal of Geosciences, Osaka City University	1954 –
Journal of the Japanese Association of Mineralogists, Petrologists and Economic Geologists	1913 –
Journal of the Meteorological Society of Japan	1882 –
Journal of Physics of the Earth	1953 –
Journal of Science, Hiroshima University, Series C: Geology and Minerology	1951 –
Kumamoto Journal of Science, Series B, Section 1: Geology	1952 – ? 8
Kumamoto Journal of Science, Geology	1972 9 –

Title of journal	Years available
Memoirs of the College of Science, Kyoto University, Series A	1923 _7_ -
Memoirs of the College of Science, Kyoto University, Series B: Biology, Geology and Mineralogy	1924 - 1967 _33_ (4)
Memoirs of the Faculty of Science, Kyoto University, Series of Geology and Mineralogy	1967 _34_ (1) -
Memoirs of the Faculty of Science, Kyoto University, Series of Physics, Astrophysics, Geophysics and Chemistry	1968 _32_ (1) -
Memoirs of the Faculty of Science, Kyushu University, Series D: Geology	1940 -
Memoirs of National Institute of Polar Research, Series A: Aeronomy	1973 (12) -
Memoirs of National Institute of Polar Research, Series B: Meteorology	1974 (2) -
Memoirs of National Institute of Polar Research, Series C: Earth Sciences (Geology, Geography, Glaciology, Seismology, Geodesy, Geochemistry)	1975 (8) -
Memoirs of National Institute of Polar Research, Special Issues	1974? -
Mineralogical Journal	1953 -
Oceanographical Magazine	1949 -
Pacific Geology	1968 -
Papers in Meteorology and Geophysics - Meteorological Research Institute	1950 -
Publications of the International Latitude Observatory of Misuzawa	1951 -
Proceedings of the Research Institute of Atmospherics, Nagoya University	1953 -
RAAG Memoirs of the Unifying Study of Basic Problems in Engineering and Physical Sciences by means of Geometry	1955 - 1968
Recent Progress of Natural Sciences in Japan	1976 _1_ -
Records of Oceanographic Work in Japan	1928 - 1975 _13_ (1)
Report of Ionosphere and Space Research in Japan	1950 -

Title of journal	Years available
Rock Magnetism and Paleogeophysics	1973
Science Reports, Niigata University, Series E: Geology and Mineralogy	1967
Science Reports, Saitama University, Series B: Biology and Earth Sciences	1951 -
Science Reports, Tohoku University, Second Series: Geology	1912 -
Science Reports, Tohoku University, Third Series: Mineralogy, Petrology and Economic Geology	1921 -
Science Reports, Tohoku University, Fifth Series: Geophysics	1949
Science Reports, Tohoku University, Seventh Series: Geography	1952 -
Science Reports, Tokyo Kyoiku Daigaku, Section C: Geology, Mineralogy and Geography	1936 -
Special Contributions, Geophysical Institute, Kyoto University	1963 - 1970 (10)
Transactions and Proceedings, Palaeontological Society of Japan	1935 -

GEOLOGY PRESENTED IN COLLECTIONS OF PAPERS

FROM THE PUBLISHED LITERATURE

Frederick Betz, Jr.

32 Park Close, Oxford OX2 8NP, England

and

A. Milo Dowden

Box 188, Stroudsburg, Pa. 18360, U.S.A.

Summary: The volume of geological literature is so great that readers
urgently need ways to identify manageable amounts of material possibly
useful for different purposes. Titles, abstracts, and indexes provide
awareness of quantities of literature, but on the whole offer limited
assistance in discerning quality. Review papers by subject specialists
are more helpful guides to literature worth examining. For direct access
to significant literature from scattered sources, organized collections
of reprints now play an important role. The scientist—editors of such
collections relieve readers of the burdens of searching through and
evaluating large masses of literature, and often also of dealing with
language barriers. The selections necessarily reflect judgments of the
editors, but are influenced by recommendations of fellow scientists con-
sulted. Some collections are planned as volumes in series that will
display the historical development of the science as a whole in a relatively compact form. The value of such reprint collections to libraries
and reference librarians requires no explanation.

Part 1 by Frederick Betz, Jr.

The vastness of scientific literature and the rapidity with which it
is growing has been discussed so often in recent years that further con-
firmation of the severe problems this imposes on users, documentalists,
and libraries or other storage places is not necessary. The geological
literature is especially troublesome. As Mather and Mason (1950)
observed in the preface to A source book in geology: "The literature of
geological science is extraordinarily voluminous, partly because of the
wide ramifications of the science itself and partly because of the tendency displayed by geologists, especially many of the eighteenth and
nineteenth century writers in this field, to set forth their ideas in
extenso and to give lengthy, detailed descriptions of the phenomena with
which they dealt." The tendency of geoscientists to be unrestrained in
their writings has not been noticeably curbed in the present century.

Since it is impossible to read all the existing literature, let
alone the newly published material, even on some highly specialized sub-

jects, the geoscientist is forced to make a decision about his commitment to reading. The findings reported by Emery and Martin (1961), Krupička (1968) and Craig (1969) show that many geoscientists forego reading not only because of the mass but of other obstacles, such as the difficulty of identifying valuable publications and of coping with language barriers. The implications of the decline in reading are regarded as serious.

The interested geoscientist committed to reading, therefore, seeks, first, awareness of possibly useful publications, second, clues to those that would actually be worth obtaining and reading thoroughly, and third, easy access to the worthwhile publications. Planned identification of useful literature by use of the familiar tools can be time-consuming and unsatisfying. Attempts to identify literature useful to an individual through bibliographies and indexes are confounded by the multiplicity of purposes that may exist in the community being addressed and the inadequacies of identifying signs. Many users, thus, fall back on casual discovery by means of scanning, browsing, and personal recommendations from colleagues who may also have come upon useful publications by chance (Price, 1967; Menzel, 1959).

An alternative for some scientists, particularly those working in a special field, is to belong to an "invisible college", which serves as a personal-communication network (Price, 1964, 1965, 1970). The members are reputed to be leaders in current research, who by exchange of communications within their groups reduce the need of the individuals to read the open literature. Papers are formally published ultimately for the permanent record. This solution does not relieve the burden of keeping up with the literature on those who are not admitted to such groups.

Reliance is often placed on abstracts and book reviews as a means of identifying useful literature relating to a specific purpose. The huge quantity of abstracts published annually has itself become an obstacle to the use of such services and the quality is so uneven that the general net value is only to provide awareness of possibly useful publications. A more effective tool is the review paper by a subject specialist, which displays progress in an area of knowledge as reflected largely in the literature. The contained references and evaluations give readers meaningful clues to selecting publications for fuller examination (Notes, 1). It may be noted that review papers are reported to be among the most widely read and important kinds of geoscientific literature (Craig, 1969). Similar service is rendered by those authors of thematic books and textbooks who introduce the reader to a sizeable amount of significant literature through quotations, comments, or simply citation. Older works of this type still generally available, in which the total literature could be and was examined, are an invaluable help in obtaining the gist of journal papers that are to be found in few libraries today.

A further step in sorting out valuable contributions to the older literature is the reprinting of difficultly or not at all available material. The historical value justifies the service that places early literature illustrating the development of the science at the disposal of modern readers. While this material will interest most those who are actively concerned with the history of geoscience, the substance of the writings often merits renewed wider consideration. Recently one assemblage of such items has been published under the heading, History of geology (Notes, 2).

Another approach to reprinting historically important communications is to present the record of the science as a whole in a sequence of excerpts. The noteworthy example cited before is A source book in geology, edited by Mather and Mason, first published in 1950 and itself reprinted more than once. In a second volume, Mather (1967) proceeded with excerpts of writings from the first half of the 20th century, which was obviously not confined to out-of-print material.

Historical value and interest are, as suggested, not the only motivations for reprinting literature. For instance, the intrinsic value of some literature, whether old or new, can make it suitable for inclusion in collections of readings for students. Selections to support specific courses of instruction in an institution and to supplement course textbooks have long been reproduced by various methods. They may represent a response to necessity (original copies not available in an accessible library or too few copies to serve the number of students) or may be a recognition of the inexperience of students in finding literature in libraries. In the past, few such collections were formally published and thereby made available for wider circulation (for an example, prepared at Yale University, see Agar et al., 1929). However, it is acknowledged or evident that many textbooks were and are developed from reading lists and informally reproduced collections tried out on students by instructors (eventually, authors).

The present discussion focuses on another kind of collection, which combines various purposes and is designed for the use of practising scientists and advanced students. It has become familiar in different sciences, no doubt because of the proliferation of literature and the consequent need for providing convenient packages of material determined to be representative and useful. It offers cohesive presentations on subjects within a science, which can be read cover to cover or selectively, or be used as reference books. Such collections can, of course, be used as

supplemental readings for courses of instruction, but are not basically
by-products of plans of instruction.

The selections may consist of entire papers or of sections of books,
monographs, etc., as well as abridgments and excerpts. Not an essential
feature, but one that identifies some collections is publication of fac-
similes, which may convey a sense of the diversity of sources to the
reader. The geoscientific selections deal with description and analysis
of features and processes, syntheses, interpretations, applications, concepts, history, and methods of investigation and communication. One
important type of information with which these collections are not directly
concerned is data. A large part of the geoscientific literature consists
predominantly of data, which in the past were re-assembled in manuals and
handbooks, and today are entered into data banks. As Price (1967) has
pointed out, the permanent (reading) record of a science does not include
"data-bank literature" and ephemeral communications. However, for valid
reasons this restriction is not rigidly observed in all collections.
Finally, the collections on a subject may be limited to writings of one
author, or from one country, or published during a stated span of time,
or to writings in one language.

In the United States more than 50 volumes of collected papers from
the geoscientific literature have been published in the last decade. A
large part of this production consists of the <u>Benchmark Papers in Geology</u>
series (Notes, 3), which is used as the example in the further comments.

A sample of the subjects treated in these volumes includes geochemistry of water and of iron, tektites, loess, glacial deposits, slope
morphology, beach features, environmental geology and geomorphology, concepts and history of sedimentary rocks and of oceanography, palynology,
and palaeobiogeography. Each volume presents an overview of the scope,
development, and factual detail of its subject. The history and nature
of the subject impose variations on the contents of the volumes. Thus,
where there is a long record of development, milestones can be recognized
in papers that are widely known and may be acknowledged classics. For
newer areas of investigation, such as different aspects of applied geoscience and interdisciplinary sciences in which geoscience participates
significantly, most of the literature may date from only the past five or
ten years. This poses a problem in forecasting the papers that will have
a lasting value at a time when the scope and detail of the subject have
not yet been established. The collections of papers may then be more an
interim overview of the subject, which should be reconsidered periodically
for revision and probable enlargement, but that does not detract from
their usefulness as displays of the state of development.

The selections reflect the personal judgment of the volume editors, but also professional recommendations and opinions obtained earlier or in the course of seeking and evaluating material, and, further, the concurrence of the series editor. It is clear that the small sample of the literature chosen to form a composite picture contains elements that could be replaced with other equally suitable ones, since few papers are truly unique. Factors other than subjective judgments also play a part in the final composition of the collections: for instance, in the series being discussed, the volumes are more or less equal in length, and, occasionally, permission to reprint a preferred paper cannot be secured.

It is fundamental to the concept of the collections that they represent the product of a survey by the editor of worldwide literature on the subject of the volume. The editor's familiarity with the subject and activity on it, prior knowledge of the important literature, and experience with leading sources give direction to the survey. However, every type of publication may be considered a possible source. The range has broadened with the development of new areas of investigation, the concurrent increase in numbers of journals and other outlets for communications, and both the willingness of and need for authors to publish in largely or wholly non-geoscientific outlets. On many topics of great current interest, journals of geoscientific societies and academies no longer serve as leading sources of important communications. Also, while attention is paid to the historical development of knowledge on the subjects of the volumes, first (original) papers are not arbitrarily selected when the information in them has been presented better in subsequent communications.

A worldwide survey introduces the problem for the editor of considering foreign-language literature. Since few geoscientists possess a reading knowledge of all the languages in which possibly valuable material may have been published, the language barrier can seriously affect the extent of coverage of sources and the selections. In recent years the increased use of English by authors in non-English-speaking countries has accelerated worldwide exchange of information, but it is a fallacy to think that all important literature is published in English or that all important foreign-language literature is promptly and correctly translated into English (Vitaliano, 1960; Emery and Martin, 1961; Routhier, 1968). As a rule, in order to accommodate to readers of the collections, papers published in foreign languages, even in the common western European languages, are presented in English versions. For the purpose, if available published translations are used, but if not, translations are specially prepared, usually by the volume editor, with a considerable, unrecognized expenditure of time and effort. When the latter has been done, some papers are thereby brought to larger audiences for the first time. In a few cases, foreign-language papers have been reprinted, but with a full English summary by the editor.

On the whole, the collections in this series are noticeably weighted with papers from the country of the volume editor, due not only to language barriers. It should be noted that review papers and collections of papers generally are characterized by a concentration on publications of the country of origin or in one language. Thus, the goal of worldwide coverage is not fully realized and it remains one to be attained in any such service. Meanwhile, some editors have succeeded in widening the perspective on knowledge of the subjects by selecting papers by authors who refer to and comment on literature that would otherwise escape the notice of many geoscientists.

Besides the selected papers, an essential feature of each volume is the text written by the editor. It consists of an introduction to the subject and its status, and commentaries distributed through the volume to introduce groups of papers. Together they form a framework to support the papers and give the rationale for their selection. In the text, the editor also draws the reader's attention to other important literature on the subject. An added value of the volumes is found in the selective bibliographies they contain, comprising the references accompanying the reproduced papers and lists of supplemental readings selected by the editors. The latter include papers that were alternate choices considered in selecting material for the collection. A consolidated author index provides access to all the titles cited in the volume. The bibliographies in some volumes must rank among the best available on the subjects anywhere.

Beyond gaining the benefit of having useful literature identified and gathered in a convenient form, the user of the collections is freed by the editors from the further time-consuming task of searching for the meaningful parts of some papers in the context of the volume by judicious abridging and excerpting. All readers of geoscientific papers will have observed that many authors stray repeatedly from their subject or feel impelled to write omnibus papers in which the subject of interest in the reprint collection is only one topic. It is not a service to the user to reprint papers in their entirety when substantial parts are not relevant. In the case of material drawn from books, there is no alternative to picking out portions, often preferably discontinuous ones. Further, as noted before, the volumes are limited in length, and therefore it is impractical to consume space unwisely when part could be allotted to other worthwhile material. Although the practice of excerpting and abridging has a long history, as in the volumes by Mather and Mason, whose observation on the

prolixity of geologic authors was cited, and is welcomed by many readers,
it is criticized adversely by some. Arguments against well-prepared cut
versions in collections of this type overlook the desirable and often un-
avoidable reasons for selective presentation, but above all the purpose
of the collections, which is not merely a re-packaging of literature.

Finally, users of these collections should recognize the editors as
themselves users (and authors) of geoscientific literature, who share
their concern about having better ways of gaining direct access to representative, significant writings on particular subjects in a manageable
form. Experienced reference librarians will recognize the collections as
a time-saving aid in serving users of literature on a growing number of
subjects with authoritative material. For libraries, the collections can
be a valuable supplement to, alternative to, or replacement for holdings
of the separate sources of the reprinted literature, especially when the
coverage of a science is offered in large sects. The collections are one
solution for some of the critical problems created by the overwhelming
mass of the literature (Notes, 4).

REFERENCES

AGAR, W.M., R.F. FLINT, and C.R. LONGWELL. 1929. Geology from original
sources: readings for students. New York: Henry Holt.

CRAIG, G.Y. 1969. Communication in Geology. Scottish Journal of Geology
5: 305-321.

EMERY, K.O., and B.D. MARTIN. 1961. Language barriers in the earth
sciences. Geotimes 6(2): 19-21.

KRUPIČKA, J. 1968. Results of the world public opinion inquiry among
earth-scientists. Věstník Ústredního ústavu geologického (Prague) 43:
249-268.

MATHER, K.F. 1967. Source book in geology, 1900-1950. Cambridge:
Harvard University Press.

MATHER, K.F. and S.L. MASON. 1950. A source book in geology. New York:
McGraw-Hill. Reprinted 1964, New York: Stechert-Hafner; 1970, with
addition to title, 1400-1900, Cambridge: Harvard University Press.

MENZEL, H. 1959. Planned and unplanned scientific communication.
Proceedings of International Conference on Scientific Communication
(Washington, D.C. 1958) 1: 199-243.

PRICE, D.J. de S. 1964. Ethics of scientific publication. Science 144:
655-657.

PRICE, D.J. de S. 1965. Networks of scientific papers. Science 149:
510-515.

PRICE, D.J. de S. 1967. Communication in science: philosophy and forecast.
In Communication in science: documentation and automation (A.J. de
REUCK and J. KNIGHT, Editors.) 199-209. London: J. & A. Churchill.

PRICE, D.J. de S. 1970. Science since Babylon (enlarged ed.). New Haven:
Yale University Press.

ROUTHIER, P. 1968. America, Europe, and Geology. <u>Geotimes</u> 13(7): 19-20.

VITALIANO, D.J. 1960. Geologists versus the foreign literature. <u>Journal of Geological Education</u> 9: 74-78.

NOTES

1. <u>Earth-Science Reviews</u> is, as the title indicates, a review-paper journal; various geoscientific journals carry review papers occasionally. Gathering of separately published review papers was undertaken by the IUGS Commission on Geological Documentation with a set of six issued from 1964 to 1970; some of them are reprints from journals.

2. This reprint series (advisory ed.: C.C. Albritton, Jr.), published by Arno Press, New York, consists of 34 books on different aspects of geoscience dating back to the 17th century, two collections of papers by Lyell and Marcou on and one collection of essays by G.W. White about North American geology.

3. This series (series ed.: R.W. Fairbridge) is published by Dowden, Hutchinson & Ross, Inc., Stroudsburg, Pa. There are also series of <u>Benchmark Papers</u> in areas of biology, chemistry, physics, and engineering.

4. It is helpful to prospective readers of such collections and to reference librarians to find a reviewer who states an understanding of the purpose and value of the collections, and displays awareness of some of the problems facing an editor, which were discussed in this paper, before presenting critical judgments. For the example, see a review of <u>Crystal form and structure</u>, C.J. Schneer, ed., 1977 (Benchmark Papers in Geology 34), by W.T. Holser (<u>American Mineralogist</u> 63: 216-217, 1978).

Part 2 by A. Milo Dowden

At any meeting of scientists and information specialists there is always talk about the growing problem of maintaining levels of user awareness, data accessability, and economic systems for both archival and new information.

The world inventory of earth science literature, according to Graham

Lea of Geosystems, is estimated to be two and a quarter to two and a half
million items and is being added to at the rate of 100,000 items each year.
Obviously, to handle the increasing quantity of information embodied in
this corpus, the new and sophisticated, even futuristic information packaging, storage and retrieval systems that have been discussed are necessary
and indeed, long overdue.

However, we may well be overlooking, or perhaps taking for granted
one of the most basic, convenient and efficient methods of packaging, preserving and communicating information - the book.

A Brief History of Reprint Publishing

Although modern scholarly reprint publishing is a twentieth century
phenomenon (camera-ready copy, photo offset, aided by institutional funding) it is actually as old or older than printing itself. Before paper,
clay tablets were reproduced extensively in Egypt and the Biblical lands
of the Middle East. Many scholars in the time of the Greek and Roman
civilizations earned their living by copying scrolls and manuscripts.
The great library at Alexandria put out catalogues of their holdings available for copying. Their salesmen covered many cities in Europe and the
Middle East. Scribal copying in the Middle Ages was heavily supported by
the church and in fact, it was the church which almost single-handedly
served as the reservoir of learned works during the depths of the Dark
Ages.

The invention of printing by moveable type around 1450 was, of
course, the major breakthrough. The business of publishing, producing
and reproducing books or scrolls, changed drastically. Publishing moved
out of the monasteries and into the universities, Oxford University Press
for example was founded in 1478. Forty-three years later in 1521, Cambridge University Press commenced production. Within seventy-five years
of the invention of the printing press, most of the important classical
works had been reprinted from scribal or handwritten editions. This was
a phenomenal achievement and included 40,000 titles with 10,000,000 copies
in print. This made manuscripts known and available to scholars throughout the world. Book publishing and reprint publishing were now firmly
established.

In the seventeenth and eighteenth centuries literary societies
spurred on the reprint business by offering works for special interest
groups. In more modern times, during both world wars of the twentieth
century there was heavy governmental sponsorship of the reprinting of
scientific and technical material. In the post World War II period in

the USA the impressive growth of colleges and universities was accompanied again by federal funding of libraries which furthered the growth of reprint publishing. Currently, there are over three hundred book publishers in the USA and Canada actively engaged in the reprint publishing business (Nemeyer, 1972).

The reprint books produced by these publishers fall into four major categories:

1. The old, rare, out of print complete book, usually facsimile reproduced, primarily for libraries and special interest groups.

2. The anthology or book of readings for instruction and student use. Selections tend to be somewhat subjective and personal.

3. The annual or "best of the year" collection of papers published by journal publishers from their own collections. The society publishers are the most active in this type of reprint publishing.

4. The archival reference reprint collection. This is the type of reprint book dealt with in this paper. In the geoscience field, the Benchmark series is probably the most extensive and widely used of any of the archival reference reprint collection publishing programmes.

A View of the Reference Reprint Collection Volume from the Publisher

My co-author has presented some interesting comments and observations about geoscience reprint publishing, and in particular, Dowden, Hutchinson & Ross' Benchmark book publishing programme from the viewpoint of the author (volume editor) and the user.

It would be of interest, I believe, to trace the development and growth of the Benchmark publishing programme from its inception to its present status. Actually, the idea for such a series began to take shape in 1969-70. This initial interest was stimulated by direct contact with the work of the American Institute of Physics.

When Charles Hutchinson and I formed our company in July 1970, the Benchmark book was still a concept but it quickly established itself as the main-stay of our publishing programme and has remained so to this day. Initially, the Benchmark programme was limited to just a few scientific disciplines, namely, geology, chemistry, electrical engineering and physics; areas which were tied to past publishing experience. It was not long before the scope of the programme was broadened to include other fields, for example the life sciences. The Benchmark concept from the beginning was that these volumes and series, edited by subject specialists (in most instances world authorities) would provide the reader with an orderly, systematic, worldwide coverage of the primary literature in their respective fields (Appendix 1).

In refining and implementing the basic idea, a number of individuals of international standing provided important ideas and suggestions, among them Rhodes Fairbridge (Columbia University), Bruce Lindsay (Brown University), John B. Thomas (Princeton University) and L.L. Langley (National Library of Medicine). Their encouragement was an essential factor in its development. Their ultimate endorsement was, of course, their willingness to accept positions as series editors.

Although a new company, with the help of these series editors it was possible to recruit world authorities to edit the very first volumes, thus setting a high standard. Some of the first volume editors in geology were Claude Albritton, John Andrews, Virgil Barnes, Donald R. Coates, Richard P. Goldthwait, C.T. Harper, Maurice Schwartz and Stanley Schumm.

The first Benchmark volumes were published in 1972 and were enthusiastically received. Of course, there were many obstacles to be overcome as would be expected for a new company launching a new and unique publishing programme. As with most reprint publishing there was the problem of securing high quality camera ready copy, no small feat when dealing with many very old and often rare texts. We soon found that the normal, xerox-type duplication process was not adequate for printing preparation. Initially, the volume editors were asked to obtain all of this reproducible copy but eventually, we established some supplemental sources, mostly libraries, for the older, more difficult material. This practice continues, and in fact, has expanded, especially in the handling of geology volumes which often contain a large proportion of very old primary material.

With seventy-five percent of the material in most Benchmark volumes coming from published archival sources, obtaining permissions from the original authors and copyright owners became a major task for both the volume editor and Dowden, Hutchinson and Ross. Since this area had been treated rather casually by some reprint publishers, it was determined from the outset to deal with this important step with special care and unquestionable business practices.

There were other developmental hurdles to be surmounted including design and format, as well as the procuring good quality but economical manufacturing. With time and experience these early problems were eventually brought into line and the Benchmark book, as it is today, evolved.

Just as Benchmark books required special and unique input by the volume and series editors and the publisher, so was the creation of a successful and efficient worldwide marketing plan. Naturally, the first step was to give the programme a name. It was our original intention to call the books and the programme "Landmarks". We soon learned, however, that this imprint was already claimed and being used by at least two other US publishers.

Confidence that these volumes would have a good reception and sale in the export markets, particularly in the developing countries rested on the need to make adequate arrangements for worldwide promotion and distribution. Fortunately, two highly respected and capable international publishers have handled the marketing of these volumes, John Wiley & Sons from 1972-1977 and Academic Press since 1977.

One burden that has handicapped the programme from the beginning and has even increased in severity is the high unit cost of producing all Benchmark volumes. As the market for these books is primarily institutional (libraries, industrial research laboratories, etc.) first printings are necessarily small, usually only fifteen hundred copies. Typical with almost all scientific, reference, short-run books, this is the major factor in determining price.

The first geology volumes carried a list price of only twenty to twenty-five dollars but general inflation, and more particularly rising permission fees have brought the average list price today to thirty dollars a volume.

One might observe that all geology books are expensive. In a 1977 survey of US publishers by the Association of American Publishers (AAP), it was found that earth science books were the highest priced of any in the technical and scientific category. The AAP study showed that the average price for all scientific and technical books was $17.48 and for geology books the average price was $25.98. Life science books were a close second at $24.66. Physical and medical science books were considerably less costly, averaging $19.51 and $16.28, respectively (Dessauer, 1978).

Earth science books are costly to produce for fairly obvious reasons: maps, half-tones and small, specialized markets. However, while most purchasers of geoscience publications have been conditioned to these somewhat astronomical price levels for new, reference monographs, understandably, they are reluctant when asked to pay similar or even higher prices for a reprint volume.

Why should a book containing seventy-five percent facsimile material cost about the same as an original, typeset work? One word answers this question - permission fees. While some reprint papers are quite old and in the public domain, most, even in the geosciences, are still under copyright and command a permission fee. These charges by the copyright owner can range from zero to as high as twenty dollars per page. Fortunately, most of the original publishers of these reprint papers (the copyright owners) in the interest of scientific communication and fair business practices, exercise a degree of restraint in setting their fees. Over the past five years, the eighteen major reprint sources for _Benchmark_ books in geology charged an average fee of $5.87 per page (Appendix 2).

The Influence of Reviewer and User Feedback on the Benchmark Programme

We have, since the first _Benchmark_ volume appeared, heeded the comments and suggestions of reviewers and users very seriously.

Special efforts were taken to improve the quality of facsimile copy, particularly, the older, poorly printed papers. For the most part, this was a learning process but the solutions have increased costs; some old and rare papers must be handled and photographed by special library personnel at fees sometimes as high as five dollars a page. Fortunately, this kind of material is usually in the public domain and not subject to permission fees.

Although the publishers and the series editors made every effort from the beginning to acquire from volume editors meaningful and substantial commentary to introduce and link the reprint papers, thie kind of input was sometimes missing. The same may be said of the subject index but that is an old story with all science book publishers. Reviewer and user feedback eventually strengthened our position in both these areas.

Another aspect relating to content that has not been overlooked by reviewers and users is the tendency of some editors to be nationalistic and even provincial in the selection of papers. Editors have been strongly encouraged to include material from all sources and all countries including papers not available in English. Funds have been made available for translation where applicable.

Current Status and Future Plans for the Benchmark Series in Geology

As of August 1978, forty-seven volumes in the geology series have been published with approximately eighty thousand copies in print. Over one hundred additional volumes are in preparation or planned. (There are eighteen active series in the _Benchmark_ publishing programme, with one hundred and sixty-five titles published.)

Tentative future plans include the publication of an annual or semi-annual author title index; some titles made available in less expensive, student editions and revised or new editions.

In conclusion the collected reprint volume in series is just one approach to coping with the growing problem of scientific information proliferation. However, of all the services which are designed to address this problem, not one makes any attempt to separate the wheat from the chaff; no value judgments are involved. For some purposes this is not a drawback but for many users, the selection process, the "weeding" out, the "boiling" down is the service that is requested. The Benchmark type reference reprint collection aims to provide the user with that service.

REFERENCES

DESSAUER, J.P. Inc. 1978. Association of American Publishers 1977 Industry Statistics. New York: Association of American Publishers.

NEMEYER, C.A. 1972. Scholarly reprint publishing in the United States. New York: Bowker.

BENCHMARK BOOKS PUBLISHING PROGRAMME, DESCRIPTION AND FEATURES

These volumes make accessible to the individual scientist and researcher, as well as to libraries, the key papers in all basic areas of scientific research, forming the very core of any science library. The papers contained in each Benchmark volume are selected by experts for contemporary impact, historical significance and scientific elegance. Each series will contain from twelve to forty or more volumes of classic and recent papers representing the landmark developments within the particular subject area of the series.

* WORLDWIDE SEARCH, REVIEW, SELECTION, DISTILLATION AND PRESENTATION OF THE PRIMARY LITERATURE
* LANDMARK PAPERS SELECTED BY WORLD AUTHORITIES
* INTRODUCTION BY VOLUME EDITORS COVERING HISTORY, STATE OF THE ART, AND FUTURE PROSPECTS OF THE FIELD
* HIGHLIGHT COMMENTARY BY THE EDITORS PROCEEDING EACH GROUP OF PAPERS
* COMPREHENSIVE SUBJECT INDEX
* MASTER AUTHOR CITATION INDEX
* TRANSLATIONS (MANY PAPERS PRESENTED IN ENGLISH FOR THE FIRST TIME)

MAJOR SOURCES FOR BENCHMARK PAPERS IN GEOLOGY

EARTH SCIENCE SOURCES	Vols.	Total pages used	Pages paid for	Fee paid	Average fee for paid page
1. Geological Society of America	28	894	0	0	0
2. University of Chicago Press	25	705	697	3,883.00	5.57
3. American Journal of Science (Yale University)	25	543	0	0	0
4. American Geophysical Union	21	519	0	0	0
5. Microforms Int. Marketing Corp.	7	499	499	2,060.00	4.12
6. Society of Economic Paleontologists and Mineralogists	11	427	69	345.00	5.00
7. Elsevier	15	368	324	3,420.00	10.55
8. American Association of Petroleum Geologists	9	326	0	0	0
9. American Association for the Advancement of Science	20	300	286	1,010.00	3.53
10. Economic Geology Publishing Co.	7	243	0	0	0
11. Geological Society of London	10	229	229	1,176.33	5.13
12. American Geological Institute	6	183	0	0	0
13. Pergamon Press	5	147	118	1,050.00	8.89
14. Springer–Verlag	10	145	74	583.54	7.88
15. Institute of British Geographers	6	122	122	665.57	5.45
16. National Research Council of Canada	7	102	72	360.00	5.00
17. Macmillan (Journals) Ltd.	16	100	24	136.42	5.68
18. Royal Geographical Society of London	7	99	86	566.70	6.58
TOTALS			2,600	15,256.56	5.87 overall average per page

GEOSCIENCE INFORMATION SOURCES AND SERVICES

FROM THE USER'S VIEWPOINT

JULIE BICHTELER

Graduate School of Library Science

The University of Texas at Austin, Austin, Texas 78712

Summary: Affecting the geoscience information user are a number of factors
inherent in the field itself: its interdisciplinary nature; the wide
variety of sources, ranging from monographs and journals to guidebooks and
aerial photographs; and the relative usefulness of older material. This
paper addresses specific concerns of the user within the context of some of
the stages in geoscience information processing and management--production
of primary sources, acquisition and storage, organization and control,
identification and location, and physical access. Problem areas include
new journal formats, escalating costs and time lags throughout the process,
inadequate indexing and abstracting, and proliferation of primary data.

Introduction

An increasing variety and number of users require geoscience information in the 1970's. Geoscientists continue to be the primary user group
with interests ranging from traditional fields to more complex, highly
interdisciplinary ones. New classes of users depend on geoscience data
and information for well-grounded decision making--city planners,
legislators/policy makers, citizens' groups, and attorneys, to name a
few. As intermediaries, librarians/information scientists not only experience many of the same frustrations as do actual users, but also face a
number of behind-the-scenes problems which indirectly affect the users, for
example, those related to acquisition and bibliographic control.

Contributing to the complexity of scientific communication within
the geosciences are a number of unique factors related to the nature of
the field itself. The relatively high value of older material in comparison with that in other areas in science and technology requires its
retention for a longer time--the contents of many geoscience publications
do not become obsolete. For example, in 1979 the American Geological
Institute will begin a project to augment the GeoRef data base by citations
from the Bibliography and Index of North American Geology from 1785 through
1960 and the Bibliography and Index of Geology Exclusive of North America
from 1933 through 1966. Such extensive retrospective coverage in bibliographic data bases is unheard of in other sciences. In addition, the
frequent necessity of relating a search to the geologic time scale or to a
geographic locality further complicates the retrieval procedure.

The user must contend with a wider variety of materials than in other scientific fields. In addition to the usual books, technical reports, proceedings and transactions, machine-readable data, dissertations and theses, and abstracts, the geoscientist requires access to maps and atlases, aerial photographs, field trip guidebooks, and informal field reports.

This paper discusses problems which are of great concern to both users of geoscience information and librarians. It concentrates on links between distribution and assimilation of information, namely, acquisition; organization and bibliographic control; identification and location; and physical access. Many of the opinions in the paper come from an informal survey among librarians and geoscientists in the United States during spring 1978, and thus, the paper reflects mainly conditions in that country.

Acquisition

In examining the process of acquisition of geoscience information, one is struck by the well-known, ominous combination of growth of the literature, increasing costs of obtaining it, and the so-called "steady-state" budgets of libraries in the 1970's. Many librarians are finding that they lack the funds to keep up.

A recent report by King Research Inc., an information research firm, provides some interesting comparisons in relative growth of the literature in nine scientific fields. The number of scholarly journals published in the U.S. in the environmental sciences, defined by King as the geological sciences, atmospheric sciences, and oceanography, is expected to increase approximately 62% during 1960-1980 (King, 1976, 2:129). Statistics for the number of papers are even more startling. In the nine scientific fields analyzed, the greatest increase in growth of papers published each year from 1960 to 1980 is projected to occur in the environmental sciences with a 239% increase! This trend is of particular concern since it is not, except in rare instances, being accompanied by a concomitant increase in serial budgets.

A recent study undertaken by the author examined the journals shown in Table 1. These titles include some of the most highly cited geoscience journals, according to Journal Citation Reports (Garfield, 1977), as well as less cited but prestigious ones considered essential in most geoscience libraries. They may be used to illustrate the increase in institutional subscription rates from 1968 to 1978. The magnitude of these increases is reflected in escalating serial budgets of academic libraries, where serials may consume one-half to two-thirds of the total acquisition funds.

Not uncommon among academic libraries these days is a no-growth serials policy in which an old subscription must be cancelled before a

TABLE 1

ANNUAL JOURNAL SUBSCRIPTION RATES FOR LIBRARIES

Journal	1968	1978
American Association of Petroleum Geologists Bulletin	30.00	50.00
American Mineralogist	15.00	50.00
Bulletin of the Seismological Society of America	12.00	45.00
Canadian Journal of Earth Sciences	6.00	35.00
Economic Geology	12.00	25.00
Geochimica et Cosmochimica Acta	75.00	155.00
Geological Magazine	12.10	64.50
Geological Society of America Bulletin	40.00	84.00
Geophysical Journal of the Royal Astronomical Society (including Memoirs)	60.00	300.00
Journal of Geology	18.00	26.00
Journal of Geophysical Research	25.00	280.00
Journal of Paleontology	15.00	50.00
Journal of Petrology	12.00	45.00
Journal of Sedimentary Petrology	18.00	34.00
Journal of the Geological Society (London)	26.00	175.00
Mineralogical Magazine	16.50	75.00
Sedimentology	26.00	92.50
Tectonophysics	35.45	466.75
Total	$454.05	$2052.75
Increase		452%

new one is placed. Other libraries are undertaking massive serial cancel-
lation projects such as that described by Swartz (1977:172) at the University of Illinois at Urbana-Champaign in which more than 2800 periodical
subscriptions were cancelled.

Compounding these woes is the variety of sources with which the
acquisitions librarians in the geosciences must deal: the usual dealers
and publishers; federal and state government agencies; local, regional,
and national societies. Since the publications of many of these issuing
agencies are not handled by the library's book dealer, separate arrangements must be made with many individual publishers.

The geoscientist depends on publications of local societies and state
geologic surveys for detailed descriptions and interpretations, as opposed
to scientists in other fields who need not contend with such geographic
aspects. Problems of awareness and availability arise from limited
numbers of copies of specific titles, frequent address changes of local
societies, and lack of national announcements of many publications. The
field trip guidebook is a typical example of such problem publications.

Librarians complain of an increasing unwillingness among geological
societies to accept standing orders. (The Geological Survey of Canada, for
example, will not accept standing orders for its papers and bulletins.)
Thus, the librarian must acquire each publication separately, a time-
consuming and inefficient proposition which results in a greater delay in
its becoming available to the user.

Acquisition of maps continues to present problems to the librarian.
When asked how they become aware of new maps, librarians list many sources,
for example, acquisition lists of large libraries, commercial map services,
catalogs of government agencies, and announcements in professional journals
such as SLA's Geography and Map Division Bulletin. A single, up-to-date,
comprehensive bibliographic source would be greatly appreciated. Geoscientists frequently have difficulty in deciding how to cite maps, and
one user suggested an inset on every map giving complete bibliographic
information in a standard format.

Organization and Bibliographic Control

The stage following acquisition, that of organization and bibliographic control, presents a number of very clearcut problems. The
greatest complaint of geoscience users is the lack of a comprehensive
abstracting service and of quality indexing in the geosciences.

Some quotations from geologists and librarians illustrate the
situation. "Our faculty members are setting up personalised indexing
services to avoid GeoRef and the Bibliography [Bibliography and Index of

Geology] because the indexing is not detailed enough for their needs."
"Searching is frustrating, as the indexing done at present is too simplified." "Compared to the old Bibliography and Index of North American
Geology from the USGS, coverage of the Bibliography is poor." "Geotitles
Weekly is too complicated and too slow for current awareness." "We need
more cumulations." "The primary problem for the user of geoscience
information is the quality, cost, and fragmentation of bibliographic
service. The Bibliography is slow, poorly indexed, and erratic in
coverage. It is priced out of reach of individuals and even some institutions." "Time lags are unacceptable."

A new complaint in 1977 is that the Bibliography no longer cumulates
the monthly citations at the end of the year; separate issues must be
bound by the library. Furthermore, the annual cumulated author and
subject indexes arrive many months late. To improve indexing of the
Bibliography users suggested a taxonomic index and more specific geographic
access including site names, formations, names of mines, etc.

In summary, the chief improvements desired in indexing services in
the geosciences appear to be: broader coverage; more narrow, in-depth
indexing; and decreased time lags. Users are wondering whether the new
management of the Bibliography will effect any of these changes. Can or
will the American Geological Institute do a better job than has the
Geological Society of America?

The lack of the kind of comprehensive abstracting service available
to most other scientific disciplines is a major factor in geoscientists'
dissatisfaction with secondary services. Many users feel that the most
perfect index cannot replace the need for abstracts. As Wheeler (1974:8)
said during hearings before the National Commission of Libraries and
Information Science in 1974, "Geoscientists miss intensely the abstracting
literature which was provided for them for so many years by the U.S.
Geological Survey." Unfortunately, the same statement can be made today.

Identification and Location
How do geoscientists identify and locate publications of interest to
them? According to King (1976, 2:312), environmental scientists depend
less on abstracts and indexes, either printed or machine readable, and
more on references from colleagues than other scientists. Table 2
summarizes the methods by which geoscientists identify references which
they eventually cite in their own publications. Table 2 appears to
confirm the opinions of geoscientists and librarians surveyed concerning
secondary services. The data are also disturbing from the point-of-view
of the enormous sums of money spent in the U.S. to create, maintain, and

TABLE 2

PAPER IDENTIFICATION METHODS IN THE ENVIRONMENTAL SCIENCES

Method of Identification	Papers Identified (%)
Direct Identifications	
Subscription Copy	28.3
Preprint or Reprint	27.0
Indirect Indentifications	
Colleague Reference	27.7
Paper, Book, or Report Reference	17.0
Abstract, Index	0.0

SOURCE: King, Statistical Indicators, 1:99.

make use of indexing services in the geosciences, both manual and computer-based. Their apparent low use contrasts sharply with the use by the physical sciences (19.8% of cited references located through indexing and abstracting services), life sciences (4%), and engineering (16.6%) (King, 1976, 1:99). One could speculate, of course, that if a comprehensive abstracting service were available, access through secondary services would be more attractive.

As we have seen from comments in the previous section, some users feel that a significant deterrent to using secondary services in the geosciences is the cost of these services. With its annual subscription rate of $750, the Bibliography seems too high-priced to many people. Perhaps it is being compared to the former Bibliography and Index of North American Geology, published by the U.S. Geological Survey in an era of more manageable serial prices. Although complaints of high cost are prevalent, one wonders if users, as well as librarians, aren't being a bit unrealistic.

A phenomenon which poses real challenges to geoscience users in the 1970's is that of increasing quantities of data. For example, as Kover and Williams (1977:41) point out:

> the biggest single problem in remote-sensing technology is the enormous flood of data that is received daily and that goes unanalyzed. The effective use of this enormous quantity of data by geologists and other environmental scientists...requires new institutional arrangements to provide an adequate level of financial support.

In a recent survey there were a large number of numeric data bases of interest to the geosciences (Luedke, Kovacs, and Fried, 1977:130).

Numeric data base systems, with their capability of data manipulation, offer services of a distinctly new kind. The U.S. Geological Survey operates many large natural resource and geological numeric data base systems, some of which are available to outside users, for example, the National Water Data Storage and Retrieval System and the National Coal Resources Data System. In addition, users have access to the Computerized Resources Information Bank and the Petroleum Data System by means of software developed for the USGS by the University of Oklahoma.

The proliferation of numeric data bases has interesting implications. How can users best be made aware of their availability, content, and use? How can standardization be encouraged which would facilitate data exchange and inter-system communication? How can error-free input best be accomplished? Raw, unevaluated data may exhibit considerable variation and inaccuracy; how can we insure that the data are properly evaluated (Luedke, et al., 1977:142)?

Physical Access

When a document has been identified as potentially useful, it still must be physically obtained. This may be accomplished through the library, the author, or the publisher, or by personal subscription or borrowing from a colleague.

Several geoscientists questioned have said that the lack of journal resources in accessible libraries is one of their chief problems. Insufficient acquisitions budgets, of course, affect the service that libraries can offer. For instance, foreign material is often held by only relatively few, large libraries. Users would naturally prefer to have material immediately at hand, and some are reluctant to understand the necessity of cooperative arrangements between libraries for providing interlibrary loans and photocopying service. Special difficulties arise in the case of geologic material: older publications may be fragile and therefore are not readily loaned or reproduced. The alternative of having personal subscriptions to current journals is becoming less feasible because of the rising subscription rates.

Many geoscientists are extremely concerned over the new microfiche format of the Geological Society of America Bulletin. They feel that the new format will not only reduce access to this important publication, but also will decrease its prestige so that the best papers will be published elsewhere. Many small libraries, as well as individual users, do not have equipment for reading fiche, much less reproducing it.

King (1976, 2:321) analyzed the methods by which environmental scientists obtain the journal papers they need. Subscription sources accounted for 57.2% of the total articles obtained, 30% of which were through library subscriptions. The remaining 42.8% were obtained through non-subscription sources (reprint, preprints, photocopies obtained from colleagues, etc.). It is interesting to note that environmental scientists use non-subscription sources more and depend on library subscriptions less than do any other group of scientists.

Summary

Requirements of geoscience information users are complicated by a greater need for older literature than in other scientific disciplines and by a wider variety of formats, some of which are difficult to locate and access. In addition, restrictions of specific geographic locations and geologic time make retrieval more complex.

The user of geoscience information faces increasing numbers of publications each year whose prices escalate at a rate greater than that of current inflation. At the same time, libraries are saddled with a diminishing financial capability to acquire these publications. Serial budgets, in particular, are becoming inadequate.

Indexing is poor; abstracts are lacking; time lags are great. Thus, to identify articles of interest, the geoscientist depends less on bibliographic services (both printed and automated) and more on references from colleagues and on preprints and reprints than other scientific users. Problems of gaining physical access, such as insufficient journal resources in local collections and interlibrary loan restrictions, result in geoscientists' use of non-subscription sources more than other scientists.

REFERENCES

GARFIELD, E. Editor. 1977. Science citation index. Vol. 13: Journal citation reports. Philadelphia: Institute for Scientific Information, 1800pp.

KING, D.W. 1976. Statistical indicators of scientific and technical communication 1960–1980. Rockville, Md.: King Research, 4 vols.

KOVER, A.N. and WILLIAMS, R.S. 1977. Remote sensing. Geotimes 22(1) : 39–41.

LUEDKE, J.A., KOVACS, G.J., and FRIED, J.B. 1977. Numeric data bases and systems. In WILLIAMS, M. Editor. Annual review of information science and technology. White Plains, N.Y.: Knowledge Industry Publications, 12 : 119–181.

SWARTZ, L.J. 1977. Serials cancellations and reinstatements at the University of Illinois library. The Serials Librarian 2 : 171–180.

WHEELER, M. 1974. Geoscientist user needs and information problems. In BICHTELER, J. Editor. Geoscience information and user needs. Washington, D.C.: National Commission on Libraries and Information Science : 3–11.

INFORMATION TRANSFER TO THE GEOSCIENTIST THROUGH THE INTERNATIONAL UNION OF GEOLOGICAL SCIENCES

ROSALIND WALCOTT

Earth and Space Sciences Library,

State University of New York at Stony Brcok, Stony Brook, New York 11794

Summary: The International Union of Geological Sciences (IUGS) promotes information transfer by distribution of formal publications, and by sponsoring conferences and symposia which lead to informal information transfer between geoscientists. Data have been collected which show that IUGS publications are well-distributed in United States geoscience libraries. Frequency of citation of IUGS publications shows that some of these publications are being used by many geoscientists. An annual bibliography of all IUGS publications is needed as these publications are often difficult to identify and obtain. IUGS fosters some useful informal contact between geoscientists, but its role could be expanded.

The International Union of Geological Sciences (IUGS) is the only comprehensive international organization in the geosciences although there are other more specialized international organizations in the field. IUGS was founded in 1961 and is now one of the three largest scientific unions in the world. Basically the work of IUGS is information transfer; information transfer between people, between different specialty groups and between different countries of the world. This information transfer is accomplished in two ways; first, by formal publication and second, by promoting informal contact at meetings, symposia, conferences, etc.

The formal publications of IUGS are numerous and varied. They include pamphlets, journals, maps, symposia volumes and monographs. The publications also have different degrees of sponsorship by IUGS. Some, like Episodes, (formerly Geological Newsletter), are closely connected with IUGS, others may have IUGS as just one of a number of sponsors. The IUGS Advisory Board for Publication stated in 1973 that symposia sponsored by IUGS should be arranged in such a way as to result either in a commercially marketable volume or preferably in a group of papers to be published in an existing journal or journals. If the nature of the symposium is such that this is not possible, then IUGS would provide financial assistance to ensure availability of these papers to the scientific

community. This means that it is often difficult to identify exactly
what is a publication of IUGS.

At the 1975 Annual Meeting of the Geoscience Information Society
(GIS), Joel Lloyd (U. S. National Academy of Sciences) suggested that
GIS study the impact of the formal publications of IUGS on United
States geoscientists. A committee of GIS then considered the suggestion and proposed two studies. First, a study of frequency of
citation of IUGS publications, and second, a study to determine
their availability in United States geoscience libraries. The first
study was made by Julie Bichteler, (University of Texas at Austin).
The report was published in the Proceedings volume of the Society
(1977). The author made the second study, (Walcott, 1978).

For the purposes of these two studies, the committee had to
prepare a list of IUGS publications. For the reasons pointed out
before, definition of what is a publication of IUGS and what is
not can be difficult. Lloyd and Bichteler compiled a selected list
of 26 publications, which can be thought of as a representative
sample of IUGS publications.

Bichteler collected data on 3-8 years of citations of 1075
conferences and symposia papers contained in the 26 selected IUGS
publications. She found that citation rates for individual papers
varied from 0-31 citations/year, with maximum rates of citation
in the second and fourth years. Rates of citation of symposia and
conference papers published in journals were more than double those
of separate conference volumes. This strongly supports the IUGS
publication policy of encouraging publication of conference papers
in journals. Citations of these IUGS papers appear in 1-89 different
journals per conference, a measure of the broad appeal and interdisciplinary nature of some of these conferences. Although citation
rates can be influenced by such factors as availability of the paper,
author's prestige, amount of interest in the topic under discussion
and rapidity of publication, Bichteler thought that the citation
data showed that IUGS publications were important to the U. S.
geoscience community.

The author's study was to determine the holdings of this same
selected list of IUGS publications in United States geoscience
libraries. In addition to determining holdings, librarians were
asked to comment on the usage and value of these publications in
their libraries. 107 of the 210 libraries polled returned the

questionnaire. All libraries owned many more conference and symposia volumes than other types of publications, such as review articles. University geoscience libraries on the average owned 64% of the symposia volumes, which is quite high. Other types of geoscience libraries, company and governmental libraries, owned about 30%. Some titles were owned by over 80% of the university libraries that responded. Company libraries showed the highest rate of owner-ship for the Lexique Stratigraphique International, (60%), and company librarians particularly commented on the usefulness of this publication.

The comments on the returned questionnaires were most interesting. Many librarians stated that it was difficult to identify IUGS publica-tions for purchase. They were grateful for the list of selected publications as it pointed out gaps in their collections, which they wished to fill. Many bemoaned the fact that there was no annual bibliography of all IUGS publications, with prices and ordering information. These comments show that IUGS does not put enough emphasis on advertisement of their publications. Many librarians like to purchase all or almost all IUGS publications, but do not have the time to track down these publications amongst the various commercial publishers used by IUGS. Other international organizations publishing material of interest to geologists pose similar problems for the librarian. It is advocated that the IUGS Advisory Board for Publication consider distribution of an annual bibliography of all their publications, together with prices and ordering information. This list should be circulated to both geoscientists and science librarians.

The majority of libraries that owned titles that are available as either separate monographs or journal issues, owned the title in journal issue form. This again supports the IUGS policy of pub-lishing in journals wherever possible, particularly if the journal is well known and widely distributed.

Most librarians commented that these publications were quite heavily used in their libraries. It can be concluded on the basis of these two studies that the formal publications of IUGS are well-received and well-distributed, at least in the United States.

The second way in which the International Union of Geological Sciences is involved in the information network of the individual geoscientist is in the promotion of informal contact at meetings, etc. The effectiveness of informal contact is hard to measure - it varies tremendously with each individual, but some of the most important information transfer accomplished in the geosciences is by word of mouth. The many committees, working groups, commissions, subcommissions and regional committees of IUGS bring together specialists from many countries to work on particular problems. Contacts are made that otherwise would not be made, and information is gathered by the individual as a result.

The author made a pilot study of the effectiveness of IUGS in promoting informal contact using the geoscience faculty of 16 members at Stony Brook as the sample. A questionnaire was circulated to them concerning their contact with the IUGS. Of course, the findings cannot be generalized, but they are of interest. For example, only 4 of the 16 had ever attended an International Geological Congress - this in a highly mobile and well-funded group of geoscientists. Although only one had ever served on a committee of IUGS, this faculty member was enthusiastic about his experiences with IUGS. Two of the 16 are involved with the International Geological Correlation Programme, and 6 of 16 were then or had been members of the international associations affiliated with IUGS. More than half had read publications from the various committees and working groups of IUGS, although not many were enthusiastic about these publications. Generally, the impact of IUGS in the informal information networks of this group of geoscientists cannot be considered high.

However, the same faculty members were much more aware of the formal publications of IUGS. Almost all had used the published proceedings of the International Geological Congresses and found them useful, although there were complaints of delay in publication and outdated material. This delay factor should be partly overcome now by the Union's decision to publish the papers of the Congresses as part of existing journals. Only one faculty member read Episodes regularly, but 6 of 16 indicated that they read it occasionally. 10 of the 16 said that they had used some of the symposia volumes of IUGS in their teaching and research.

Most of the faculty were not enthusiastic about the organizations of IUGS, perhaps because their research is more experimentally oriented than regionally oriented. Although IUGS provides an organizational framework for informal contact between geoscientists, the effectiveness of this contact depends to a large extent on the

individuals making up the various groups. Some groups work extremely
well, others languish from lack of interest and leadership. It would
be interesting to know how geoscientists from "isolated" countries
rate the importance of IUGS as a means of establishing contact with
other geoscientists. Because of the large number of geoscientists
working in the United States and Canada, considerable informal
contact is easily achieved on the national level without resort to
international organizations. Therefore it is possible that North
American geoscientists underestimate the value of contacts that
might be made through IUGS. It would be of interest to learn of
the importance of IUGS in fostering contact between scientists who
are not part of academia. The patterns of communication are different for academic scientists and governmental and company
geoscientists.

It seems to me that the best that the IUGS can do in promoting
informal information transfer is to continue its present sponsorship
of the International Geological Congresses and the various specialized
symposia and conferences, and also to continue to provide an
organization which allows contact in many different ways for many
different groups. I feel that the Union recognizes its members'
needs in this regard, even if the members do not always realize it.
From information from faculty members at Stony Brook, those
geoscientists who are involved in Union activities are enthusiastic
about their involvement.

Conclusions:

The investigations show that IUGS has an effective publishing
program. The formal publications of IUGS are well-distributed in
United States geoscience libraries, despite justified complaints from
some geoscience librarians that these publications can be difficult
to identify and purchase. The publications would definitely be more
widely distributed if they were better publicized. This could be
achieved by the circulation of an annual bibliography of all IUGS
publications to the geoscience community.

Frequency of citation of IUGS publications shows that many of
these publications are being used widely by geoscientists. They
also clearly demonstrate that the wider the distribution of the
particular paper and the easier its availability, the better are
its chances of being read and cited. Wide distribution seems to
be best achieved by using established journals for the publication

of conference and symposium papers.

There is evidence to show that IUGS promotes some informal contact between geoscientists. However, the author considers that IUGS could do much more in this role. The concept of "working group" is a useful one, but IUGS should make an effort to interest the most effective organizers in heading the various groups and committees of IUGS as a good way to ensure the greatest participation in and the highest productivity of these groups.

REFERENCES

BICHTELER, JULIE. 1977 Publications of the International Union of Geological Sciences: their influence on U. S. geoscientists Geoscience Information Society, Proceedings 7 : 1-16

WALCOTT, Rosalind. 1978 A survey of the holdings of a sample of International Union of Geological Sciences (IUGS) publications in selected U. S. geoscience libraries Geoscience Information Society, Proceedings 8 : 96-110

STATUS OF INFORMATION EDUCATION FOR GEOSCIENTISTS

IN THE UNITED STATES AND CANADA

DIANE C. PARKER

Lockwood Memorial Library

State University of New York, Buffalo, New York 14260

Summary: In the past decade, teaching information retrieval skills has
become more prevalent throughout the United States and Canada. In 1978
a survey of 465 colleges and universities showed that both libraries and
academic departments are engaged in teaching geoscience students how to
find information in their field. While there is widespread activity and
great variety in methods used, no one really knows how many students receive this kind of instruction. Survey results suggest the proportion is
small compared with the total number of geology students.

Introduction

Geoscience information systems and the body of geoscience literature
are vast in scope and complexity. A great deal of information is available to the proficient user who knows how to find it, but how are these
information retrieval skills taught? The information specialist has a
role to play in this part of the educational process. In addition to
developing the technology to control and disseminate information, geoscience information specialists are teaching people how to use information
systems. In the United States and Canada, this sort of teaching, commonly
called library instruction, is being fostered by several library associations. For instance, within the American Library Association, two
separate groups were formed in June 1977 to promote library instruction.
One of them, the Bibliographic Instruction Section of the Association of
College and Research Libraries, specializes in academic libraries. In
addition, agencies such as the Council on Library Resources have provided
thousands of dollars in grants to libraries developing instruction programs (Association of Research Libraries, 1977). This report reviews what
is being done to teach geoscience students how to find information in
their field.

Methodology

From January to March 1978 a survey was conducted of librarians and
professors at colleges and universities which grant degrees in the geosciences. It was assumed that these institutions would have the primary
concern for the education of geoscientists. The American Geological

Institute's <u>Directory of Geoscience Departments in the United States and Canada</u> (1977) was used to identify 465 institutions which grant degrees in the geosciences. In this directory, the term 'geoscience' is defined broadly to include not only geology and other earth sciences, but also mining, planetary science, environmental sciences and oceanography. The survey was in two parts, with two questionnaires.

The first questionnaire was called the "Preliminary 1978 Survey of Information Education for Geoscience Students in the United States and Canada". Its purpose was to identify persons in both academic departments and libraries who are actively involved with providing library instruction for geoscience students. This questionnaire also included questions about the institutions' geoscience collections and the availability of computer searching of bibliographic data bases. 403 institutions (87%) responded.

The second questionnaire, called the "1978 Survey of Formal Library Instruction Programs for Geoscience Students in the United States and Canada", was sent to 279 institutions which had some sort of systematic or formal program of library instruction for geoscience students. It was designed to gather specific information about the library instruction offered. 167 institutions (60%) replied.

Results From the First Questionnaire

A few of the 403 institutions said they had dropped their geoscience programs. Of the 395 usable replies, 352 (89%) reported that library instruction is offered by the library, and 81 (21%) reported that it is offered by a geoscience department.

Institutions were asked to indicate the kinds of library instruction offered. Where such instruction is given in libraries, 322 institutions reported that it is available informally to individual students on request, and 293 said it was given as a normal part of reference/information service to individuals. For the purpose of evaluating survey results, this kind of informal teaching was not considered part of a consciously developed program. More formal efforts are summarized in Table 1.

TABLE 1: Types of Instruction Offered by Libraries

Type of Instruction	Number of Institutions Reporting	%
Tours of facilities	287	73
Lectures in library	195	49
Lectures in classroom	109	28
Library exercise	72	18

TABLE 1:   (continued)

Type of Instruction	Number of Institutions Reporting	%
Single-session; specific course needs	212	54
Multi-session sequence	23	6
Semester-long course	21	5
Course-related instruction	33	8

Fewer geoscience departments also give library instruction. Their informal instruction includes referring students to specific books or authors (64 institutions) and referring students to indexes, abstracts and other bibliographic research tools (70 institutions). More formal efforts by geoscience departments are listed in Table 2.

TABLE 2:   Types of Instruction Offered by Geoscience Departments

Type of Instruction	Number of Institutions Reporting	%
Instructors bring individual students to library	40	10
Instructors bring class to library; demonstrate publications	38	10
Instructors bring class to library; demonstrate catalogs & library tools	31	8
Formal course taught by member of the department	10	3

Information about the institutions' geoscience library collections was requested to learn if the occurrence of library instruction is related to the complexity of facilities. Can library systems which have multiple facilities to staff also afford staff for library instruction? The survey shows a substantial amount of involvement by all types of libraries, but library instruction is somewhat less prevalent in systems which have at least one geoscience library with additional geoscience collections in other libraries. The difference is not great, about 7%, but it increases for types of library instruction which require heavier staff involvement (credit courses, seminars, etc.). Occurrence of library instruction programs in various types of libraries is given in Table 3.

TABLE 3:   Instruction Offered by Various Types of Libraries

Institutions Where the Library's Geoscience Collection is:	Number of Respondants	Number Giving Library Instruction	% Giving Library Instruction
A separate geoscience library	34	31	91
At least one geoscience library with additional collections as part of a science and/or general library	35	29	83
Part of a science and/or a general library	326	293	90

Formal library instruction programs and computer searching of bibliographic data bases are services which have become more common in academic libraries in the last ten years. How likely is it for an institution to have both computer searching and library instruction? Of the 235 institutions which have computer data-base searching, 219 (93%) also offer library instruction. Some characteristics of these computer services are listed in Table 4:

TABLE 4:   Computer Services in Academic Libraries

Service	Number of Institutions Reporting	%
By mail request only	39	17
On-line	176	75
Off-line	100	43
Requester present for search	137	58
Requester not present for search	129	55
Search done in a geoscience library	11	5
Search done in a non-geoscience library	147	63
GeoRef searched	125	53

Survey results indicate that formal library instruction by library staff is more likely to occur at universities with graduate programs than at four-year colleges. Conversely, formal library instruction at colleges is more likely to be given by staff of a geoscience department.

When asked what kinds of library instruction were offered, most respondents took the word 'offered' quite literally. In some cases, students don't take advantage of what is available. Often, librarians

are willing to give instruction, but only when someone requests it; they
don't always take the initiative. Nevertheless, it's clear that most
libraries answering the survey felt a responsibility for teaching students
how to use library materials.

Results From the Second Questionnaire

The second questionnaire was sent to 279 institutions which indicated in the first questionnaire that they offered some kind of formal
library instruction for geoscience students. 167 (60%) institutions
replied. Information was requested regarding type of instruction, course
credit given, level and number of students taught, course content and
types of teaching materials.

A variety of instruction is offered with tours of library facilities
being the most common and semester-long courses with weekly sessions being
least common. Types of instruction are summarized in Table 5.

TABLE 5: Library Instruction for Geoscience Students

Type of Instruction	Number of Institutions Reporting	% of 167
Tours of facilities	139	83
Lectures in library	102	61
Library exercise	51	31
Single session; specific course needs	109	65
One or more sessions as part of a general program of library instruction	34	20
One or more sessions as part of a geoscience course	39	23
Multi-session sequence (mini-course, seminars, etc.)	15	9
Semester-long course with weekly sessions	11	23
Course-related instruction (library component planned by library & geoscience department staff)	23	14

Only eleven institutions reported having a semester-long course, and six
of them actually were designed for the general student body, not geoscience students. Only five semester-long courses of library instruction
were specifically for geoscience students. All of them are given for
academic credit by a geoscience department. Two are taught by professors,
and three are taught by librarians.

In most institutions students are not given academic credit for library instruction. Where credit is granted, it can be from a geoscience department, a library school, an education department, an English department, etc. Rarely is credit granted by a library, since most institutions view libraries as service units which cannot grant academic credit.

What proportion of geoscience students in the United States and Canada actually receive library instruction? The latest data available on enrollment is for the 1975/76 academic year. According to a report prepared by the American Geological Institute (1977) for the U.S. Geological Survey, the total 1975/76 enrollment for all geoscience fields, including freshmen through doctoral candidates, was 26,987. How many of these students received library instruction? The answer seems to be that we don't really know. Most institutions could not provide statistics for that year, and many which did could not separate out their geoscience students from other students. The figures given here are edited and probably inflated. A generous estimate is that 1,935 geoscience students received library instruction in 1975/76, which is just 7.2% of the geoscience student population.

Course content is varied but usually includes general library materials and methodology as well as geoscience information sources. Usually the focus is on bibliographic forms such as geoscience indexes and handbooks (113 institutions) rather than the literature of sub-disciplines such as mineralogy and geochemistry (45 institutions). Instruction on use of government documents is common (111 institutions). Use of maps (66 institutions) is taught almost as frequently as use of general science information sources (68 institutions). Computer bibliographic searching (38 institutions) is taught least often.

Exercises and worksheets are the most frequently use printed instructional materials (40 institutions); reading lists (34 institutions) and course outlines (27 institutions) are next most common. Tests used to determine a student's level of ability before instruction begins are seldom used (4 institutions), and tests or examinations to determine level of achievement are used only slightly more often (12 institutions). Evaluation forms for the students' use in giving a course critique are relatively uncommon (21 institutions). 46 institutions reported the use of audio-visual equipment, usually an overhead projector or slides.

At the end of the second questionnaire 71 respondents indicated they would be willing to have their institutions listed in a directory of library instruction programs for geoscience students, and 52 said they would be willing to help develop a model course syllabus.

Conclusions

The survey indicates there is interest in information education for geoscience students throughout the United States and Canada. However, there are few well-developed instruction programs. Most of the interest is among librarians, but librarians usually do not take the initiative in promoting information education. Again and again returns carried phrases like "given at the instructor's request." One librarian said, "Our instructional program is initiated by requests from the faculty." Unfortunately, geoscience professors do not always recognize a need for information education. As one geology department chairman said, "Students are expected to know how to use a library." And so, students sit in the middle, receiving little if any library instruction.

The responsibility for giving and promoting information education needs to be shared by information specialists and geoscience faculties. Information specialists know bibliographic systems and information control; they could be teaching students to use information sources. However, it's the professors who give students a sense of what is important and what should be learned. There is a world of difference between the professor who says students are expected to know how to use a library and the professor who teaches a course on information sources. There needs to be understanding support and a strong partnership between the two professions.

To summarize, work in geoscience information education has begun, and if the trend of the last few years continues, it may be a common part of geoscience education in the 1980's.

REFERENCES

AMERICAN GEOLOGICAL INSTITUTE. 1977. Directory of geoscience departments in the United States and Canada, 1977-1978. 16th ed. Washington, D.C.: The Institute, 168pp.

AMERICAN GEOLOGICAL INSTITUTE. 1977. Student enrollment in geoscience departments, 1975-1976. Washington, D.C.: The Institute, 88pp.

ASSOCIATION OF RESEARCH LIBRARIES. 1977. Library use instruction in academic and research libraries. ARL management supplement. 5:1, 6pp.

SOME PRACTICAL APPROACHES TO GEOSCIENCE
INFORMATION PROBLEMS OF DEVELOPING COUNTRIES

A. R. BERGER

Association of Geoscientists for International Development,

Geology Department, Memorial University, St. John's, Newfoundland, Canada

and

V. RICALDI

Executive Coordinator, Association of Geoscientists for International

Development, c/o Alirio Bellizzia, Ministerio de Energia y Minas,

Centro Simon Bolivar, Torre Norto Piso 19, Caracas, Venezuela

Summary: Despite the great variation in conditions from one developing
country to another, there are certain features of the state of geoscience
information which are common throughout the spectrum from the poorer to
the richer ones. For example, the proportion of scientific literature
emanating from developing countries or the numbers of geoscientists from
such areas in attendance at international meetings such as the present
one suggests that even these are measures of the stage of development.
In the following we discuss in summary form some of the common geoscience
information problems that characterize many developing nations, some of
the practical approaches that the Association of Geoscientists for International Development (AGID) has developed in an attempt to ease them, and
some of the possible new directions that could be taken.

Some Common Geoscience Information Problems in Developing Countries

Governments and institutions invariably assign low priorities to and/or
cannot afford adequate geoscience information services. The results of
this are:

> poor services (if any) for the collection, storage and dissemination
> of such information;
>
> badly staffed and equipped libraries, often not open to outside
> workers;
>
> general non-availability of selective dissemination of information
> (SDI) and other information service taken for granted in more
> developed countries;
>
> the paucity and often prohibitive cost of duplicating facilities; and
>
> large accumulations of unpublished documentation, much of which is
> labelled "confidential", partly because this avoids the problem of
> dissemination.

Individual geoscientists though placing a high priority on information matters face a number of problems:

they are often ignorant of, or are unable to gain access, to standard geoscience information services and publications;

are apt to be too dependent on standard textbooks, especially if trained entirely at home;

have little encouragement to publish their own work and even then face impossibly high page and offprint charges;

have little tradition of scientific writing and are not always able to produce an internationally-acceptable paper despite the possibly high quality of their research; and

commonly possess little sense of scientific community and are too often isolated from the mainstream of geoscience as expressed at conferences, meetings, etc.

Some Current Approaches

AGID, an international non-governmental organization with over 1200 members in 90 countries, is dedicated to the building of a viable geoscience community in developing countries and to increasing the contribution of the geosciences to rational and equitable development. AGID works on a scientist-to-scientist level and has established a number of activities relevant to geoscience information:

the organization or regional and international conferences, workshops, training courses (groundwater research, mineral development policies, mineral resource management, geoscience education, geochemistry, mineral exploration in tropical rainforests, etc.) designed expressly for developing countries by geoscientists from such nations;

a programme of low-cost publications available to developing countries free upon request or for which payment can be made on an in-kind basis. These include a quarterly newsletter, a series of directories of geoscientists and organizations, and a Report Series on the Geosciences in Development;

the recycling of surplus books and journals to developing country libraries otherwise unable to gain access to past and current literature. Nearly 3000 items have been distributed in recent years to libraries in Peru, Bolivia, Nigeria, Sri Lanka, Uganda, Zambia, and Argentina. There is a continuing need for donated material. Please search your files and bookshelves and contact AGID;

a Geoscience Information Service for requests for scientific and

technical purposes. AGID acts as a clearing-house and referral
centre whereby geoscientists from developing countries can gain
access to SDI services and to information centres and networks.
Recent requests have been concerned with raw materials for glass-
sand industries, methods of analysis for metals appropriate to
tropical conditions, geochemical exploration in laterite terrains,
small-scale mining, etc.;
an international survey of the groundwater information capabilities
and needs of developing countries has just been completed. This was
based on a Bibliography of Developing Countries Groundwater compiled
from extensive searching of various on-line and other bibliographic
services (GEOREF, GEODE, NTIS, etc.). For 1970-1976 there were
approximately 1500 citations to published material. The survey suggests that this represents no more than about 2-5% of the total published groundwater literature on all countries for this period, and
more importantly only about 5-10% of the estimated volume of unpublished documentation in developing countries. Clearly the existing
international bibliographic services are monitoring only a very small
fraction of the total groundwater literature on Africa, Asia and
Latin America; and
the preparation of a list of about 300 key books and journals for
geoscience libraries in developing countries, which is intended as a
guide to the fundamental literature.

Some Continuing Needs and Possible New Directions

A major need is for the establishment of regional and national information
centres backed up by offprint services and biased first to local requirements and then to a wider network system to enable the rapid international
flow of geoscience information among developing countries. As a first
step towards such a system, inventories (bibliographies) of geoscience
information in such areas should be made by local geoscientists who are
in the best position to judge their own needs. AGID's strategy in this
regard is to identify active and interested geoscientists to determine
their information needs and priorities, to assist them in planning appropriate projects, and to secure the necessary funding. A pilot study is
now underway in Bolivia with the long-term aim of establishing a groundwater information system for that country and a possible prototype for
other developing countries.

There should be active promotion of publications written by developing country geoscientists, possible through some kind of revolving fund which can provide support in return for a portion of sales revenues.

Consideration should be given to the establishment of journals of translation of key papers written in major languages other than English, especially into French and Spanish (eg an International Geology Review in Spanish).

There should be a much wider availability of geoscience information services such as the existing on-line systems. Formal or informal links between geoscience societies and institutions in all countries could provide a mechanism.

Perhaps the most fundamental need is for new initiatives to encourage the more effective usage of geoscience information by developing country geoscientists, ranging perhaps from short courses on report and map preparation, to the use of existing international information systems, and to the design of appropriate information systems.